"Bring Me Men…" Brought Women

"Bring Me Men…" Brought Women

Marching with the First Female Cadets at the U.S. Air Force Academy

KATHLEEN UTLEY KORNAHRENS

McFarland & Company, Inc., Publishers
Jefferson, North Carolina

ISBN (print) 978-1-4766-9059-9
ISBN (ebook) 978-1-4766-4866-8

LIBRARY OF CONGRESS AND BRITISH LIBRARY
CATALOGUING DATA ARE AVAILABLE

Library of Congress Control Number 2023000385

© 2023 Kathleen Utley Kornahrens. All rights reserved

No part of this book may be reproduced or transmitted in any form or by any means, electronic or mechanical, including photocopying or recording, or by any information storage and retrieval system, without permission in writing from the publisher.

Cadets marching through the Bring Me Men Ramp
(courtesy of USAFA Clark Special Collections, McDermott Library,
Photography Section); the author during Parents' Weekend,
September 1976, at lunch in Mitchell Hall with our families
(author's photograph)

Printed in the United States of America

*McFarland & Company, Inc., Publishers
Box 611, Jefferson, North Carolina 28640
www.mcfarlandpub.com*

To Fred
my love, my Falcon buddy, my best friend,
and the inspiring embodiment of character,
honor and integrity
And to
Mary, Anne and John.
You awe me with your successes,
your dedication and your caring hearts.
Your lives are my inspiration.

Acknowledgments

I can't image my life without Fred, and without him, this book would never have been reborn from the old files and stories. He patiently encouraged, researched, cooked, poured wine, did laundry, all so I could write. Beyond thanks—my love forever.

Gail Benjamin Colvin, thank you for your loving friendship from the compassionate hug after the pugil stick competition to the encouragement, wise counsel and fun ever since.

To all the 80's Ladies: together, we did it!

A huge Falcon cheer for all USAFA brothers who stood with us, appearing not in the papers, maybe, but in our hearts.

Thanks to my dad, a pillar of strength and understanding, for believing in me and for taking me to the library so many times to get stacks of books to devour. I wish you were here to read this one.

Mom, I treasure the picture of you in your WAVE uniform. Thank you for your service and for your example of caring for the less fortunate. One hundred two and still going strong.

A giant hug and much love to Mary for starting the whole process by asking why, for your living example of an empowered and empowering woman and for caring always.

Annie, you are an amazing inspiration. You see every person in life's story and work to make them heard. Thank you for your "words" of wisdom and your motivation.

To John, USAFA '16, a warrior in the skies but more importantly for fairness and compassion. For your service and your encouragement, thank you, love and hugs. NIC.

Thanks to my sister MaryAnn Hofmann who listened and offered tea, blue cheese and cara caras for sustenance during the process.

Much gratitude to my sibs, Pook, Mike, Pat and Barb, who wrote letters that saw me through many tough days of cadet time.

My Niner sisters: the six who began and graduated. Go Niners, beat 'em!

To all the Stalagers, especially John, Ron, Bob, Freddy, Pat, Dave, thanks for being my brothers in the pain and in the fun.

Andrew Ward and Annie K., thank you for all your time spent reading with keen, intelligent and very compassionate eyes.

Rob Wilson, Hank Wilborn and Ron Sheffield: You are all amazing officers who set the standard for leadership and caring, in and out of uniform.

Much thanks to all USAFA McDermott library staff then, in between and now, especially Command Historian Brian Laslie and Ruth Kindreich and Joel Hebert, PhD in lieu of staff of Special Collections for many hours with USAFA histories, oral histories and heavy boxes of files and for a quiet moment in the Gimbel Library to focus and get perspective.

To the AOG staff (special thanks to Amanda Hess and Gary Howe), thank you for the support to me and all in the Long Blue Line.

To the teachers at The Loft, Minneapolis, who first called me a writer. I so appreciate the encouragement.

To all the helpful hands and minds at McFarland for encouragement and understanding to bring my story to readers.

Table of Contents

Acknowledgments — vii
Preface — 1
Women Graduates of the Class of 1980 — 3
Prologue — 5

1. Nominee — 11
2. Basic — 24
3. Invader — 40
4. Pugilist — 61
5. Beanie — 78
6. Guinea Pigs — 85
7. Doolie — 91
8. NRA Shooter — 107
9. My Darling Daughter — 118
10. Falcon Love — 133
11. Black Panther — 147
12. Hell Week — 155
13. American Pig — 162
14. Three Degree — 176
15. Two Degree — 184

16. Señorita	193
17. Firstie	196
18. Graduate	212
Epilogue	223
Chapter Notes	233
Bibliography	239
Index	245

Preface

  I lived this story and then revisited it as I wrote it down to document my part in the history. It is not a universal story of the first women at the Air Force Academy. Every man and woman who participated or watched has their own. Through letters written to me while I was a cadet and letters I wrote, I remembered the situations and my feelings from the time. I even kept journals of events including writings on the back of my survival training map. My story then grew from questions and thoughts about the removal of USAFA's "Bring Me Men" words. It grew too as other men and women from my class shared with me their stories, some similar and some very different from mine. I added representative dialogue of events, not exact words, to help describe the very unique USAFA environment. My pride in our class's accomplishment grows as I grow older, and I am thrilled to offer my slice of this history to readers.

Women Graduates of the Class of 1980

(Graduation Ceremony Program, Class of 1980,
United States Air Force Academy, 28 May 1980)

Theresa Armbruster	Allison Hilsman	Catherine McKenzie
Carrie Banwell	Bonnie Houchen	Rosemarie McLeish
Karen Barland	Mary Hughes	Theresa Meyer
Deanne Barnett	Mary Jeffrey	Ann Moore
Gail Benjamin	Kathleen Johnson	Kathleen Moriarty
Nancy Berger	Karen Jorgensen	Mary Morse
Kathleen Bishop	Betsy Joviak	Diane Moyer
Andrea Bopp	Lucille Juhas	Lynn Nagahiro
Virginia Caine	Ellen Kincaid	Nancy Newberry
Marjorie Clark	Jan Knocke	Karen Novak
Karen Cole	Gwendolyn Knuckles	Karen O'Hair
Kathleen Conley	Susan Kohut	Karen Olson
Jeanna Cumnock	Doris Krampe	Marianne Owens
Mary Daley	Mary Lademan	Beverly Plosa
Sandra Darula	Deborah LaFrombois	Michele Pompili
Peggy Davis	Lisa Lambert	Chris Reasner
Margaret Dennis	Karen Lange	Julie Richards
Susan Desjardins	Laurel Langmade	Lorraine Roemish
Allene Dowden	Janet Libby	Patricia Ryan
Debra Dubbe	Phyllis Loving	Bonnie Schaefer
Holly Emrick	Donna Lundquist	Debra Senn
Linda Garcia	Karen Lusk	Tanya Senz
Marjorie Glazier	Dorothy Mahaffy	Pamela Simonitsch
Diana Green	Nancy Mariano	Laurie Slavec
Joanne Guretsky	Anne Martin	Ricki Smith
Susan Helms	Susan Mashiko	Dara Synder
Sue Henke	Susan McAdoo	Mary Jo Synder

Women Graduates of the Class of 1980

Emi Takashima
Janet Therianos
Paula Thornhill
Susan Timmons
Maureen Tritle
Beverly Turman

Eva Turner
Andrea Ungaschick
Kathleen Utley
June Van Horn
Mary VanValkenburg
Peggy Walker

Sandra Waples
Deborah Wilcock
Karen Wilhelm
Lenora Wong

Prologue

The thin, arid air of the Colorado Rockies burned my lungs. It felt familiar, though, almost friendly. *I know this air,* I thought. *It was mine once, a long time ago.* My skin wasn't so complacent. It was dry and itchy. I remembered that feeling too. Colorado air sucks the moisture from skin, leaving leathery toughness on the outside at least. My heart felt itchy, too, distressed as I thought about the awful sex scandal that was keeping the United States Air Force Academy (USAFA) in the headlines. Somehow, the dishonor of my alma mater seemed to implicate me as well. All graduates probably felt some of the same shame that this icon of integrity and honor was being forced to admit in public arenas that its environment was being considered one of sexual harassment and assault. As a graduate, as one of a few female graduates, as one of the first female graduates, the whole scenario made me want to wash my hands again, which would just chap my skin more, not rid me of association with the home of this sex scandal. *How can I love this place, its atmosphere, and be so ashamed all at the same time? How can I hold so close to my heart a mostly male, predominantly macho-cultured institution? How can I bring my own daughters here to see this military academy when it is in the headlines for its mistreatment of women?* But without logical explanation, this spring break I was here to embrace USAFA. This place, my alma mater, "mother of my soul," and me, the mother of these three children in the back seat, were going to have a face-to-face, and the children were going to be the witnesses.

"Keep a latrine on your radar screen," I said to my husband, Fred, also an academy grad. We didn't just revert to the military nomenclature when we passed through the south gate. We always used these military words and phrases with each other. The three in the back seat of the rental car knew some of them too. They could sing cadences about C-130 aircraft and echo other lines from marching tempos that Fred and I remembered from years ago. "C-130 rolling down the strip…. Airborne Mama gonna take a little trip…." This breeding ground for Air Force officers was not only our alma mater, but it was also the beginning of our family stories and many of the words that found their way into our lexicon. Like latrine. Fred wrinkled his brow to remember where the closest latrine would be along this route.

The mountain views were beautiful. The Air Force Academy had been my home, our home, for four years. This should have been a comforting homecoming for my family and me. *Darn it! I owned this place years ago. It took me four hard years, but I conquered these roads in combat boots, carrying an M-1 rifle.* The anxiety I felt on this return was distinctive too. Unlike my cadet years, I was not returning to exams and inspections. I was not worried about physical fitness testing that usually ended near a trash can with the dry heaves and a beet-red face. This time, I was a returning graduate from a historic class. I wiped the palms of my hands on my "good black pants" that I put on after a week in jeans and ski pants to present the correct reverence for this institution. My gripping headache from earlier in the week had disappeared thanks to my acclimation to the over-mile-high altitude and a daily dose of Tylenol. My sunglasses protected me from the glare reflected off the snow on the grandiose Rampart Range against which this aluminum university nestled, but my eyes still throbbed as I attempted to absorb each view.

My head pivoted west, catching sight of Falcon Stadium as Fred started retelling our cadet woes of the mandatory football games and march-ons. "We were required to go to all games," he explained. "We had to wear our full uniforms and sit facing the sun. I remember games when I felt like a chicken on a rotisserie and other games when it was so cold, we wet a blanket in the latrine so it could freeze to make a windbreak. Some officer reamed us out for it. Something about 'conduct unbecoming a cadet.' We were shocked. We thought it was so inventive." Fred bemoaned the frequently defeated Falcons of the 1976–80 era to our three young football fans in the back seat. "They rarely won," he said, "but we were dutifully, always in the stands."

Our heads also turned to the east. I had to tell our kids about the airfield right there on the academy grounds where Fred started his flying career, and I gained a hard-earned glider pilot's certificate. I didn't recount the details of the long-ago Saturday morning, there by the tarmac, that found me awaiting the return of my instructor pilot. He was also my friend, a partner at the laundromat, a great adviser about the "way of USAFA," and a buddy with a car who shared a ride out of the gate for an ice cream or a burger. But he did not return from that last flight. Ever. Blinking, I could see his face in my thoughts. After his glider spiraled down into a Kansas field, leaving me crushed emotionally, I was afraid to go back into the sky. For several nights in the dorm, my summer roommates kept the overhead lights on to deny the shadows a hold over my dreams. Finally, some need for closure nudged me aloft for a final solo checkout ride. I got a solo certificate, which I tucked into my flight suit pocket along with a picture of that friend that had accompanied me to altitude. I never flew a glider again, but my children knew I had once. They reveled in the recognition of their mom's accomplishment without the distress of the whole story.

As our car traveled closer to the cadet area, the median of the road still smelled of sewage although it was only a ghost odor. Brown grass filled the road's concrete centers. The color surprised me. My mind pictured the vibrant green lawn fed by the semi-treated wastewater as a younger me drove past these medians toward the dorms. Appearances were everything at USAFA. I remembered one Saturday morning hoping a parade would be canceled due to heavy rains and quagmire conditions. The noisy thumping as helicopters from Fort Carson hovered over the field to dry the stage for our performance for some visiting dignitaries put the harsh reality of the parade right back on the schedule.

Many have asked me what I thought of the grandeur of the academy only to be shocked with my absurd answer that I barely saw it for the first year I was there. From day one, screaming upperclassmen demanded that our eyes be locked to the front with absolutely "no gazing around." Today, however, I was gazing in every direction. Sometimes my eyes slid 90 degrees to the left to catch the profile of my 17-year-old daughter Mary. What did her own eyes betray of her feelings? Could she ever know how my 17-year-old self felt seeing it all for the first time as she did now? "The academy scares me," she once told me. "I guess I've heard too many war stories." I certainly had told her how scared I was, too scared to speak or even to blink sometimes. I had no real idea then that I was engaging this institution and its all-male history in battle for a place for me and for women forever. I knew that I would be among the first, but I was more scared about the rigor and the harshness than about the momentous change the presence of female cadets signaled.

But Mary knew of some of the happy times too. She knew I met her dad here. She had seen pictures of some of our crazy dressing up for squadron meetings. All three of our children had heard us talking about night spirit missions, cadet speak for pranks, and wild cadet ski trips. I hoped they heard pride in our voices when we talked about USAFA. They had seen our sabers mounted like trophies on the wall above our fireplace. They knew we were back here to show off, not to condemn. I wanted to share more stories of the fun but could not find them in my scrambled thoughts. Maybe once the butterflies of 27 years ago quieted, I would resurrect some positive tales of our time here.

The giant Pegasus statue stood in front of Doolittle Hall. "That used to stand behind Arnold Hall," I announced.

"Oh, yeah, Hercules's horse, from Disney," the back seaters explained to one another.

Fred's fingers touched my hand on the center console, both of us knowing but not telling the rest of the Pegasus story. Some lore is appropriate for certain audiences only. Appropriate for 17-year-olds, 14-ers maybe, but not nine-year-old boys. Not this day. The story of male cadets traditionally taking girlfriends out behind Arnold's, the recreation hall, to touch Pegasus and telling them the cadet fable that the giant winged stone horse would fly away if touched by a virgin. No. If fully told, our nine-year-old son would need an understanding of virgin, the mythological history of Pegasus and a strong admonition that he should never treat a girlfriend like that. Better left unsaid for now.

The car's final climb up the winding road into the rocky walls took my breath away as had many runs up this same path back in 1976. Running was part of the core curriculum of academy life. I would never miss that part of my cadet experience, but I was proud that I had run.

"We are almost there, so straighten up," I warned our USAFA offspring.

Why? Their eyes asked, but their mouths didn't. They knew that Dad and Mom were more tense than usual. They were disciplined enough to obey and question later.

The steel-and-stone visitors center where we were to meet our guide seemed deserted. "Mom?" the kids questioned. "Who would be here? We saw the signs at the south gate: *USAFA Closed to Visitors.*" Post-9/11 security and the Iraq war sealed the base to all but a select few.

But true to military-fashioned rigid schedules, we found the center's doors unlocked, only the staff within the building. We had no real time to shop, but we were a few minutes early. We scurried through the display about the history of the Air Force Academy. It began with photos of the Class of '59 in Denver, the excavation of the site in Colorado Springs and the construction of the chapel. Knowing we were nearing the arranged meeting time with our guide, I took in one more section. A framed picture caught my eye. "Bring Me Men…" in giant silver letters on a photographic image of the portal just yards from reminded me of where my academy story started. Our class, including the first 157 women ever, marched under that portal and those words on the first day we arrived and back beneath when we graduated. *That's really why I am here. The kids need to see all of this, but more so, I need to see it again.*

My military mindset dominated. "Let's not be late," I said and hurried the group outside to my waiting classmate Ginny Caine Tonneson. I recognized her right away, even though we were both older and not the super-fit young women who marched here in our cadet years. Even though we rarely got together, the "80's Ladies" bonds that got us through the trial of our years here kept our group tight. Ginny was working for the Association of Graduates and provided us access to the areas of the academy now off-limits. I was keen to take it all in.

During the hike over the hillock to the cadet area, Ginny dropped the bomb. "They took down the 'Bring Me Men...' sign yesterday, you know." No, I didn't know. Fred didn't know. We had just spent two generator-only days isolated from war and other news in the mountain retreat of a family friend. We knew the academy was boiling with a sexual assault scandal. Both Congress and the public were proposing solutions to an institution they struggled to fully understand. I initially thought it would all blow over. I hoped the stories were not true or that the cadets dismissed over the rapes and rape allegations were all caught, all gone, and the women all miraculously healed. I played ostrich, not wanting to think about the environment that I thought I already knew, warts and all. I hadn't even known about the sign's absence but immediately felt saddened by its removal. Our son, John, overheard and added, "No fair. That's the place Mom is writing about."

Why did they? How encompassing is this scandal to have sent such momentous shock waves to the academy and public leaders? What did the sign have to do with sexual assault? Along with 156 other young women, I fought and cried to be one of those mentioned "Men." I never let the sign deter me. I accepted, no, I ignored it, dealing with the in-my-face issues of survival. In my shock now, I didn't know how to feel.

That shock, not the cold, biting wind, numbed me. I kept repeating, "I can't believe it" until after seeing the dorms and dining hall we walked to the edge of the library like doubting Thomases to witness the now-nameless ramp. Annie, my sensitive, free-spirited young teen, scooted closer to hold my hand. Small bits of dusty accumulation from almost 40 years left the shadowy outline of some of the letters. "Bri_g M__" was all we could decipher of the beginning words of the Sam Walter Foss poem: "Bring me men to match my mountains, / Bring me men to match my plains, / Men with empires in their purpose / And new eras in their brains."[1]

"They did it while the cadets were away on spring break," Ginny contributed. "There will be an uproar when they all return." *The similar uproar of my stomach,* I thought. Times four thousand.

It all left me hollow as our footsteps echoed on the deserted terrazzo. John skipped ahead following the linear marble strips that were our required paths for a whole year when we started as cadets. He moved along the cool gray-and-white stone blocks because he wanted to, because he still had that childlike interest in stepping on one block after another, avoiding the imaginary perils of the spaces beyond the paths. How could he know that he would dislike these blocks later in life? How could he know that he would be honored with his name chiseled on a broken piece of one of these same blocks?

Mary hung back with her dad, so Ginny and I could compare thoughts. I wanted to hold Mary's hand. I wanted her to understand my grief for a lost signpost that was part of the traditions of this academy that had I joined in my own way. Mary was a very strong and silent 17. She kept her chin up and her feelings to herself. But she watched me. She noticed the slump of my shoulders. I know she did.

I tried to stand tall for the rest of the tour. *This beautiful but flawed institution is my alma mater.* This was my first home on my own. I was one of the first women to wear a United States Air Force Academy uniform. Nothing would ever change that. This is where I was tormented for being "a female" but remained so proud of my womanhood. This is where men called me "bitch" inches from my face, then later whispered sweet words close to my ears. This is the place where I gained the confidence for the life that I eventually would lead. Where I met the love of my life.

We moved out of the cadet area. I was still shell-shocked over the transformation. We went back into the gift shop to purchase the photograph of the ramp, now a collector's item. I bought magnets and mugs and decks of cards all showing the American flag waving and cadets marching through the portal that proclaimed, "Bring Me Men…."

Mary looked over my shoulder at my purchases and asked, "Why would you care, Mom?"

1

Nominee

"The Air Force Academy act prescribes that our institution shall be for the training of 'selected young men.' If you can get the law changed, I shall be delighted to take in young ladies as well."

—General Hubert R. Harmon,
First Academy Superintendent,
November 16, 1954

It was the picture that the nation, and probably some of the world, saw. The photo graced the front pages of many U.S. newspapers on June 29, 1976. One headline shouted in bold, dark letters: "Sex Barrier Falls as Women Enter Air Force Academy."[1]

The pleased look on the young woman's face in the photo was remarkably reassuring. She seemed thrilled with the prospect of her upcoming adventure. Her blond hair blew in the wind, giving her a look of freedom, an air of untamability. Dressed in a '70s A-line skirt and top, stripes converging like chevrons in the center, and holding a small suitcase in her hand, she was caught giving the cameras one last smile over her shoulder as she headed with an air of confidence up the large ramp that led to her new home—the United States Air Force Academy. With blue skies and warm sunlight framing her form, the whole captured image seemed idyllic. The only flaw in the entire scene were the words that loomed large over the girl's slender figure—the giant silver letters that were inscribed on the top of the ramp spelled "Bring Me Men…."

That picture could have been of me. I was there. But I wasn't blond or smiling. My recently cut red hair rebelled and curled toward every compass point. My mouth was set in a grimace, my teeth hidden behind tight, bloodless lips. My suitcase pulled heavy at my right arm, unused to weight, and drew my posture askew. The tiny calico blue jumper I wore looked childish, not the stylish ensemble of a '70s-era feminist. I never stood apart, out in the open like she did to be caught alone in a camera lens, hardly the strong, confident, new female cadet the photographers wanted.

Shutters clicked all around me, but no photographers pointed their lenses in my direction.

But I was one of the 157 women who invaded the all-male bastion of the United States Air Force Academy that day. We came from all walks of life and came in all sizes and shapes.

* * *

I was scared, sweaty scared, as I sat on the airplane bound for Colorado. I had flown before but never alone. I was leaving my small central Florida home to become one of the first female cadets at the United States Air Force Academy. *Crazy,* I thought. Heading for the Air Force Academy and being scared of flying made an idiotic picture in my mind. I feared more than just the flying.

The wide black tarmac of the Tampa airport radiated its heat up through the metal body of the airplane. The air-conditioning was working, but with my seat belt on, according to the directions given by the stewardess, my arms weren't long enough to reach the tiny nozzle and unscrew its opening. I let the rivulets of sweat reach the waistband of my skirt before I quickly unfastened the belt as quietly as I could, stood, opened the vent and quickly rebelted. Unaimed, the air hit the top of my head. Slowly, like mercury dropping in a glass thermometer, the air cooling my mess of red curls transmitted chill down to my armpits to stem the flow of perspiration.

My small, hard-sided Samsonite suitcase was stowed in the baggage compartment. It was a high school graduation gift from my parents in the latest color of the '70s: harvest gold. Many kitchen appliances of the day matched it. The suitcase was not full. Paperwork from the academy included strict guidelines about what we could bring along to this new life. I had packed the black low quarter shoes that my parents helped me mail order from Sears. They were like band shoes, only dull black. I had worn them around the house to break them in but never outside. They were odd-looking shoes, and even if I never had stylish jeans or peasant blouses, I did have a nice pair of platform heels that I wore to school with most outfits. I wasn't going to be caught out in these granny shoes until I claimed the uniform to match. Also in the suitcase was a hair dryer that would lie unused in a locked room for many months, a small bag of toiletries that I would not see for as long and the paperwork that told us not to bring anything else. It was a light bag. If the baggage handlers pulled my bag from the luggage cart with the same strength they gave the larger, overstuffed bags, it would fly over the conveyer belt to the dark cave waiting in the belly of the jet. It was a Braniff Airlines plane, painted one of the colors of the tropical fruits that grew in this sunshiny state.

1. Nominee

I wonder if there is a way to get the bag back and get off before takeoff? Maybe I should have taken my sister up on her offer to have me along for her vacation ride to Colorado to spend that last week before entering the academy seeing some sights and cruising the byways. I was tempted. Not only was I afraid to fly alone, but I was even more afraid to arrive alone. But I didn't accept. I was most nervous to arrive a minute late.

It is funny to think that the military mindset of "on time to all things" was already germinating in my psyche. I knew, even then, that the academy was going to be serious business. I think I also wanted to delay as long as possible the no-turn-back point, the actual commitment to attend the academy. I already signed all the papers, but up until the last few weeks, it was only an exciting adventure somewhere on the horizon. People from town, people I hardly knew, congratulated me wherever I went. Relatives from all over the map called to gush about how wonderful it was that I had been accepted with the first class of women to enter one of the service academies. The local paper wrote up my acceptance as front-page news, out of only a few pages but still right there with the first view. My father's chest grew by inches each time the subject was broached. All around, things were positive. Frankly, it was an easy thrill to be singled out as one of the first group of women to be accepted at the Air Force Academy. But so far, I hadn't had to actually do anything.

As the engines of the jet began to whirr, I pressed my forehead against the double-paned Plexiglas of the oval window. Through the darkened windows of the terminal, I could see Mom and Dad waving a final salute of goodbye. Dad, seeming always taller than his 68 inches, was definitely smiling. Mom stood close, purse hanging from her arm. She was probably praying. That's what she does.

I'm off. Goodbye, brief life in central Florida. I would not really miss it. I didn't know what I would miss, and I certainly didn't know what I was getting into.

* * *

Later, the rumble of the engines and the rocking of the fuselage lulled me away from my fear of the skies and of the future. The past seemed to fill my mind. *Why am I even doing this? What curious chain of events brought me to this point in time, on this airplane, alone and unsure? What brought the academy to this momentous change to even accept me?*

I know now that there was another young woman, 19-year-old Jone French, who jabbed at the all-male bastion years before I did. She was a freshman at the University of Oklahoma and gained a nomination to the academy from Oklahoma senator Henry Bellmon in March 1970. Jone French said she was "willing to take her case to the supreme court"[2] and said she "hopes girls from every state will apply to the service academies so that congress will be forced to make some provision for them." She attended the Dean's Ball with a second-class cadet from Aurora, Colorado. She attended the Dean's Ball, but her nomination did not bring her to USAFA as a cadet. The academy held firm on the requirement that cadets be "male citizens of the United States ... with a minimum height of 5'6" (waiverable to 5'4")."[3]

I've been told that it didn't seem to be her true desire to attend the academy. She would do it, but she wasn't really behind the idea. There were undercurrents of thought that she had been put up to the request by others. It seemed some in the academy administrative offices agreed that she wasn't passionate about attending.

It was many years after Jone French's stab at acceptance, years filled with many conversations and much consternation, before the doors of the academy were open to women. Although women had served and were serving in the military, the process to consider women at the service academies began in earnest with the congressional passing of the proposed Equal Rights Amendment in 1972. The possibility of ratification by the required number of states brought the conversation about women being integrated into the military academies into the main forums. High-ranking officials of the military services discussed whether women had a legitimate and equal place in the military of the United States. It seemed an almost certainty that coeducational military academies would become reality sometime in the near future.[4]

In 1973, a suit was filed on behalf of four women who had been nominated by their state representative to the Naval and Air Force Academies. The judge in the case ruled against the women, saying, "the academies had a legitimate government interest" in excluding women, justified by the fact that academies were to train combat leaders and women were at the time precluded from serving in combat positions.[5]

In 1972, Title IX could have possibly forced open the doors to all of the United States Military Academy, the U.S. Naval Academy and the U.S. Air Force Academy to allow women to be cadets or midshipmen. Title IX of the Education Amendments of 1972 "prohibits discrimination based on sex in education programs and activities that receive federal financial assistance,"[6] except the amendment excluded "educational institutions training individuals for military services or merchant marine."[7] The academies did not have to comply.

Title IX, "the one law that was on the books specifically dealing with sex discrimination in education[,] would not be of any help in the plight of women to gain admission to any of the service academies. This was compounded by the fact that the equal rights amendment seemed distant in its process of ratification."[8]

The Air Force Academy thought a change would eventually come. Congress was considering the issue and would have the ultimate decision on the question of women attending military academies, so USAFA began to plan.

The planning exposed many questions. How many women should be accepted? Would there be a dual-track system, or would the women be incorporated into the existing training protocols of USAFA? How would facilities be adapted and how would women fare physiologically to the expectations of athletics and military training?

Early in the evolution of the plan concern was expressed by Gen. Robert J. Dixon, deputy chief of staff for personnel, in May 1972 that "only those modifications essential to accommodate the basic physiological and strength differences between men and women should be made. Female cadets should meet the same, or comparable replacement, graduation requirements as the male cadets."[9]

The fact that planning for women as cadets had begun was reported by the Associated Press in August 1972. There was also some early discussion about how forward thinking and innovative the Air Force Academy would be and would seem if they proposed accepting women before the change was mandated and forced on them.[10]

While all these questions were being debated in the academy's planning committee, Congress began the deliberations to consider the change in the law. Ironically, a true pioneer in women's aviation made a surprising statement to Congress during the debates "Concerning the Admission of women to the service academies: Thursday July 18, 1974; House of representatives subcommittee no. 2 of the Committee of Armed Services Washington D.C.: Miss Jacqueline Cochran."

In introducing herself to the committee, she said, "I have had a rather unique career as a woman. I was in England during the Battle of Britain, I flew a bomber across the North Atlantic, and was able to help cure a problem in Canada that was going on among the pilots that had been drafted away from many of the areas that were not in war at the time." She had 15,000 hours of flying, been shot at over the north Atlantic and crossed the Atlantic several times during World War II.

She began her testimony saying, "A woman can do anything that a man can do that does not require brawn and I think a certain type of exposure." But, she continued, "it is my firm and considered opinion that women should not be permitted to enroll in the Air Force Academy. They are pretty difficult, women are, I know, I have dealt with them my whole life." She went on, "The basic reason why I hold this opinion is that the academies are for the purpose of teaching combat to their students." And finally and so surprisingly, she said, "A woman's primary function in life is to get married, maintain a home, and raise a family."[11]

She did later advocate for "equal pay for equal production" but not for women at USAFA. The greater irony is that as cadets, we marched by a display case of memorabilia and the uniform of Jacqueline Cochran, an aviation pioneer.

"Finally, in May 1975," one researcher described, "Congressman Stratton placed an amendment to the Fiscal Year 1976 Defense Authorization Bill, directing the secretaries of the military services to admit women to the academies."[12] The signing of Public Law 94–106 on October 7, 1975, by President Gerald Ford allowed women through the previous "Men Only" doors to the military academies—including me to USAFA.

* * *

I don't even really know why I attempted it. I was 17. The whys were not as large as the coulds and the imaginings for me back then. Later, much later, after we graduated, I heard other women's stories of their decisions to attend USAFA. I was awed by their research and dedication. I was humbled by how I accidentally ended up in their midst. But in 1976, I was thankfully naive.

No grandiose tradition or academic-based factors entered into my decision to choose the Air Force Academy. Sure, my father was a career Army man; my mother was a Navy veteran. They finished the Sunday crossword puzzles without the help of the dictionary, but neither of my parents had gone to college. There was no family or any tradition that could have welcomed me to the academy. There had been no women ahead of me. No one, least of all me, knew what to expect with the arrival of the first 157 women to cross the threshold.

I do remember a high school classmate being an unwitting influence on my choice. He was not even a boyfriend or any kind of romantic interest, just a guy from my high school class who was semi-nice to me. His interest exposed me to the existence of the academy. He decided in his junior year of high school to apply to the Air Force Academy. He always wanted to fly. His father was career Air Force. Mike held some true motivation and was genuinely gung-ho about the place, the curriculum and the career that would follow.

I remember the other inauspicious event that attached to me like a tiny tick. One high school afternoon, I edged by the health/science teacher to find my seat in his room. This teacher wasn't even the science type. The mysticism that surrounds a chemical mixing or potential explosion gives a science teacher an edge in the awe and respect category. This man was on my schedule, my health teacher, droning on during class about food, nutrition and cleanliness. In those days, there were not even the titillating discussions about drugs and sex to make that one class in the nine-week session interesting. But that day in 1975, he was the one who informed me of the momentous change in the admissions to allow women into the service academies.

Just months earlier, at the academy, "plans for the recruitment of women candidates shifted into high gear with President Ford's signing of the Defense authorization Procurement Bill which included the provision for admission of women. The Academy printing plant printed 50,000 copies of new brochures *The Air Force Academy: An introduction* and *USAF Academy: Guide for Women Students and High School Counselors* and had 50,000 additional brochures printed in Denver. News releases were sent out to thousands of high schools and to Liaison Officers to drum up women recruits."[13] My science teacher had probably seen one of the newly minted pamphlets. "Why don't you try it, Kathy?" he suggested. I'm not sure he was even serious. It may have been a passing comment as I entered his class that day. I was a newcomer to the school. I didn't know he knew me. But he said it, and it poked me and left its venom. Or antidote.

I had other options. I could stay in our one-stoplight central Florida town to become an anomaly to everyone there. I really confused townspeople from the beginning, moving into town as I did in the middle of my junior year from Costa Rica, where I had lived for the last two years. Costa Rica, a country that sounded like Puerto Rico. It couldn't be, though, because Puerto Rico was surely an island and we had driven up from Costa Rica. Our car's Pensionado license plates clearly spelled out Costa, not Puerto. I also confounded everyone being able to speak fluent Spanish but not being a migrant worker. I was comfortable, though, with the migrant workers and their children. They were outsiders too. I hung out mostly with them at the beginning. Local kids began to ask how I learned to speak English so well.

Muddying the waters further, I arrived in town with no known background, the kind most southerners spew out as if it were all a part of a long oral history permanently attached to the end of your last name. *You know, Billy Joe Stiles, the son of Junior Stiles and Mavis Branch of the William Branches whose granddaddy started at the Co-Cola plant here right before the big flood, you know the one....*

That place in central Florida was where I had lived, but it was not my home. Staying there was not a good option. I knew I could never really fit in or be able to explain away my weird past. Spanish-speaking Kathy Utley, from somewhere else, whose grandparents no one knew, just didn't fit the local norms.

My other choice was a partial scholarship at the college turned university from which my oldest sister graduated. The majority of the professors were still nuns and most of the enrollment was still female. It was by far the safer bet, a proven track driven by my older, dependable sister. But I didn't want to follow her. I didn't want to follow anyone. I wanted my own track.

I knew I wanted something different, something for me. It seemed like a good idea at the time. Or it seemed like a way to escape the shadow of my sister and brothers, a way to forge a path that would be mine alone. It was a way that my mother, my father and even my brother had experienced in a uniform in the service of the country, but the uniform and the experience would still be uniquely mine.

USAFA was different. I had no idea how different, but I did know it was a chance to break out of the ordinary. I didn't consider myself a raging feminist. In all honesty, I would have loved to be one of the girls pictured in our high school girl-a-month calendar. "Miss Any Month." Even my high school exchange sister from Chile made the calendar cut. I would have settled for homecoming something, even princess twice removed during high school, but was never even considered. I subscribed to no bra-burning craziness, but somehow (without my convictions consciously attached to it) Helen Reddy's "I Am Woman" was like a mantra to me. I liked the song. I sang out the lyrics loudly, strongly and horribly out of tune everywhere I went around the house. "Yes, I am wise, but it's wisdom born of pain" came from a 16-year-old girl who had never genuinely experienced pain, apart from the tightening of my orthodontic braces once a month. I drove the Utley household crazy. My family vowed to permanently seal my lips with glue if one more syllable of one more word of the lyrics from that song reached their ears. I started humming.

My mother had worn a uniform. Her experience might have been a motivation for me to try military life. She was in fact one of the first Navy WAVES who, as the acronym (Women Accepted for Volunteer Emergency Service) suggests, were only accepted as temporary volunteers. The WAVES were established on July 30, 1942, as the U.S. Navy's corps of female members. During World War II, some 100,000 WAVES served in a wide variety of capacities, ranging from performing essential clerical duties to serving as instructors for male pilots in training.[14] Mom joined in March 1943. She told stories of yelling by the men who trained them and of being marched wearing skirts into mud and left to stand and sink up to their ankles. She joined to get away, hoping to see the world. She was made a storekeeper second class and managed salvage in Norman, Oklahoma. I don't remember her telling the stories until after I had mine to tell, but I had seen pictures of her in uniform when I was a child. She looked so glamorous, the sepia tones of the image softening the reality of the experience.

If my father, a retired Army career man, had ended up as a

chauvinistic old-timer with outdated ideas, I might not have dared to try the whole military scene. He did sport the old-school Army crew cut all his life. His physique reflected the life he had led: strong chiseled features and ramrod-straight posture. His commanding voice echoed through my ears and down my spine. "When I say jump, you ask, 'How high?'" He was tough and fair and strong. Later, a classmate would excitedly tell his wife that my father had been a general in the Army. No, I corrected him. Allison's dad was the general. Mine stood like a general, walked like a general, led like a general, but he had retired with colonel insignia on his shoulder, a general only in his demeanor and in my heart. I respected all that he embodied.

The Army that he served in World War II and Korea did not include many women. He was an artillery soldier in World War II, fighting his way battle by battle, bullet by bullet up through Italy to eventually see Mussolini hanging in the piazza in Milan. He knew a whole different picture of the fighting man than I could have portrayed with my tiny frame and curly red mop of hair, no matter what weapon I ended up carrying. As an artillery officer in Korea, after attending the "90-day wonder" school of officer training, he again saw war. He might have been tough, chauvinistic and hardened to the softening of the military with female genes. Passed over later in his career for the sake of two West Point grads, "ring knockers" as he called them, he disdained those service academy graduates who felt that their academy commissions as officers entitled them to any special consideration. My application to USAFA might have been a nail in his coffin—a cause for disowning me as daughter and heir. But the opposite happened. He was my number one supporter.

Although extremely strict with our behavior, he put no limits on our dreams. When we were little, we played Mass all the time. We flattened white bread in the pages of the *Encyclopedia Americana,* and we cut out round hosts with my father's shot glass. With a beach towel draped over our shoulders as the vestments, we all got to play the priest, girls and boys. "You can be anything you want to be," we often heard from Dad. Turning off our one black-and-white set in the middle of a show, he'd demand, "Don't believe anything you hear on TV." "Be your own person." "Do your own thing." "Non illegitemus carborendum: NIC," Latin, according to him, for "Don't let the bastards grind you down." Not the talk that you would expect from a 26-year military veteran, born in 1918, a go-by-the-regulations kind of man, but those were the words we heard as part of our daily routine. Was I somehow mentally prepared by my upbringing to be an accidental pioneer at the academy?

Regardless, I had not applied to the academy to blaze a trail for women everywhere. I was not trying to cement some long-standing feminist belief for other women in the world. I just wanted my own place, my own adventure. Thanks to that support at home, I believed very strongly in me. I was going because I could. I was falling into an opportunity to prove myself to be something that no one else had ever been.

*　*　*

My stomach still queasy, my hands still clutching the motion sickness bag discreetly against my thigh but at the ready, I thought I would never relax until the wheels of that wild colored flying machine retouched the earth in Colorado. But my eyelids drooped. Again, my body relaxed despite the fear in my mind. The hum of the engines and the bounce of the seat slowly replaced the future fear with the current opportunity to rest.

No wonder I was worn out. The whole process of applying, getting a nomination and finally being accepted was so involved. It was an elaborate step-by-step formula. By the time I finished the paperwork and requirements, I was so invested that it would have been hard to be anything but eager.

I kept the process to myself as long as I could. No need to get Mom and Dad overly enthusiastic before I knew what this was all about. I got info from my classmate Mike and began on the long list of squares to be filled for USAFA acceptance.

First, on my parents' green Olivetti typewriter, I wrote to apply for a nomination. The smudged carbon copy dated December 12, 1975, began, *My father served 26 years in the U.S. Army, and he has bestowed upon me a sense of pride in the United States of America. I would like very much to be in the service of my country.* Wildly, years later my daughter using a laptop would type very similar terms in her college essays citing the tragedy of September 11 as a reigniting of her sense of patriotism.

Because of those years that my father invested in the Army, I was eligible to skip the congressional nomination requirements section. I got to go straight to the president. "There are unlimited nominations, but only 100 appointments available annually in this category. Vacancies allocated to the President of the United States have been reserved for children of career military personnel."[15] This section applied to me. I wonder if that was the academy's way of honoring those who served or a manner of seeding the classes with some who might know a little more about the military way of life. Maybe those of us from regimented backgrounds were meant to be a stabilizing force.

That was, however, just step one. A nomination was not an appointment. I was getting winded from the application.

I tackled the further list of involved and sometimes tedious necessities before I could be considered for acceptance. I still had to get a

Physical Aptitude Examination (PAE), according to the brochure: a "test of strength, agility, speed, and endurance. The examination consists of the following five events: pull-ups (men)/flexed-arm hang (women), standing long jump, modified basketball throw, push-ups, and 300-yard shuttle run."[16]

What I didn't know when I read the requirement was how many Air Force Academy committee hours had gone into even the one parenthetical differentiation of flexed-arm hang verses pull-ups. The Special Commandant's Committee of Integration (no, Admission) of Females (no, Women) pondered every nuance of terminology and requirement to get that one sentence in the brochure. As directed, the committee only considered separate standards if "justifiable based on physiological differences" and kept to a minimum.[17] In fact, the brochure I read in 1976 was one of the more than 100,000 newly minted, using models as the female cadets. After the law was signed, with only eight months before our arrival, the academy's admissions department hastily sent these brochures to high school counselors, to college and public libraries, and to ROTC and recruiting offices to encourage women to apply.[18] Maybe that is how my science teacher got the word. Other women found out through teachers, parents still in the military or family friends.

To comply with the PAE requirement, I rode over to McDill Air Force Base, Tampa, with my school buddy Mike. There were a few other nominees waiting when we finally found our way to the base gym. I don't remember seeing any other girls. Standing around awkwardly in groups of two or three, we acted skittish and afraid of the test and of each other. Some chatted about SAT scores and congressional support, but I kept by Mike's side, quietly. Wearing the short polyester shorts from my PE class, I threw the basketball from my knees on the polished gym floor. While the men did pull-ups, I hung from the metal bar feeling the skin of my palms pinch and burn until my biceps gave out. I ran back and forth from block to block, squeaking my low Adidas sneakers at every reverse. I did push-ups and springy jumps to prove my athletic prowess. It was all over in half an hour, but the nerves, gone haywire that morning, did not quit their jumping and pushing and squeaking for many hours. I was finished with another check mark on the requirements list.

Next to do, a physical at a DODMERB (Department of Defense Medical Examination Review Board) or authorized site. Time to get a checkup and get used to all the acronyms. The process was not simple or quick.

Finally, I interviewed with the local liaison officer, Major Carl Arant. He wore his Air Force uniform to the house and brought a booklet for my parents and me to read. There were women pictured in some of the photos, stand-ins to portray the future. He showed a movie so he must have brought a reel, projector and screen to the house. In black and white I saw the academy for the first time. I can't remember what he asked me or I him. I didn't know what to ask.

Having all the squares filled left me in anxious waiting. I had so much invested in the whole procedure just to get accepted. Even though I first considered the academy almost casually, I became zealous.

When my official academy letter arrived dated 30 April 1976, I was deemed "Qualified, No Vacancy," a kind of wait listing for a slot. More than 1,700 women had applied for the 157 spaces.[19] It seemed like there were better choices ahead of me in line. Unless one of them refused, I was not going to be one of the pioneers at USAFA.

Reading the form letter, with the anxious eyes of my parents and younger sister on me, I was relieved and saddened at the same time. The tears welled in my eyes, but my disappointment wasn't enough for them to spill onto my freckled cheeks. The pressure and mayhem to complete all the requirements had built up my hopes. The excitement about the possibility to take my own path felt right, but I was nervous about the significance of the decision to attempt the academy. Did I jump right into the fire looking for a change and ignore other less dangerous options? I had tried. I could still look for other colleges that would challenge me. At least now I could relax knowing I had given this USAFA option my best shot. I just had to figure out where I was going instead and how to get there.

One week later, before other ideas and options had a chance to cement themselves, the second notice arrived postmarked USAF Academy, Colorado, rescinding the first and offering me a position in the Class of '80, the first coed class of the United States Air Force Academy. Some other woman or some other man had said no. I didn't know at the time how the selection process worked, but now, with enough sour grapes in my attitude to have softened the blow of the first rejection, I was not sure I wanted to go. The investment of applying had been eaten away by that waitlist semi-rejection. *I just might not do this.*

I left the house. I wanted to distance myself from the decision and the papers waiting to be signed. I walked to our neighbor's and dangled my legs in the tepid water of her pool. The sparkling blue mesmerized my eyes. I tried to focus my thoughts on other things. I let my mind waver with the flashing reflections of the ripples around my legs. I jumped in and let the air bubble out of my lungs as I stretched flat on the bottom of the pool. I stalled and thought hard about thinking about nothing. I saw the black-and-white images of the mountains and buildings and the airplanes slipping in and out of frames. I saw my father's face filled with pride. I felt the pinching of my palms on the metal bar and the ache in my knees slamming hard onto the basketball court. I surfaced for air and sank again to the bottom to think. I thought of the town where I didn't fit in. I remembered my picture in the paper. I remembered my brother in pictures from the jungles of Vietnam and my mother's sepia image in uniform. In my 17-year-old mind, anxious to build its own memories, I could find no reason not to go.

After I signed the acceptance with the black military-style pen that my father carried in his breast pocket, even all those years retired, he enveloped me in a steeling hug and whispered, "*I am so proud*" in my ear.

2

Basic

"If you can't get them to salute when they should salute
and wear the clothes you tell them to wear,
how are you going to get them to die for their country?"
—George S. Patton

Landing in Colorado Springs, June 27, 1976, brought my head upright with the bumping of the gear on this high western tarmac. I could have been at an Air Force academy built in Alton, Illinois, or in Geneva, Wisconsin. A team of planners including General Spaatz and General H.R. Harmon, who would become the first academy superintendent, the president of the state university of Iowa, the vice president of the Hearst Corporation and famous aviator Charles A. Lindbergh considered "580 proposed locations in 45 different states." But Geneva and Alton displayed some public unrest over the proposal to be the home of the future academy. Signs of "go to Colorado" in Lake Geneva did not bode well for the academy being accepted or built there. Colorado Springs embraced the plan and lobbied for the Air Force's academy to be built there.[1] Charles Lindbergh wanted to take a small plane up to survey the site from the air. The Pine Valley airport manager did not want to let the unrecognized pilot take up one of his planes. He asked Lindbergh if he knew how to fly, then asked him to show a pilot's license. Lindbergh showed all his pilot licenses from various countries to the embarrassed manager who authorized Lindbergh's flight. Lindbergh declared the area acceptable for flight training, and in 1954 President Eisenhower okayed Colorado Springs as the home of the future Air Force Academy. The *Denver Post* quoted Senator Ed Johnson's reaction: "This is the greatest thing that has happened to Colorado since Pikes Peak was discovered by Zebulon Pike."[2]

Inside the Colorado Springs terminal, a tall blond man, Chris Young, and his lovely dark-haired wife, Nina, walked right over to me with a question in their voices. "Kathy?" Chris, an academy graduate originally from central Florida, had been linked to me through my liaison officer. They didn't know me at all. We quietly walked through the terminal with my harvest-gold, hard-sided suitcase swinging like an empty shopping bag on Chris's strong arm. I didn't ask Chris much about the academy, and he didn't volunteer much. We talked about Florida and Colorado and a bit about my family. Chris told me just a little of what awaited me after tomorrow's dawn. I slept fitfully at their home that last civilian night. Years later, my own son would spend the night with an unknown family the night before he began his USAFA adventure. He didn't want us to bring him out and drop him off even in the age of helicopter parenting. *It will be hard enough to say goodbye,* he told me, *but to say goodbye and begin the shock of training before recovering from the goodbye will add too much emotion to the day.* My parents couldn't afford the airfare to come with me to Colorado, and the parenting style of the '70s would not have encouraged such attention to offspring. I made my way with strangers to begin the journey.

I heard once that when you arrive at a new location, you bring everything you have been with you. What had I been? I had been a female, a girl, mostly, a tomboyish girl. Once I had to pretend to be a boy to start a business with my brother. "Odd Jobs, Inc. We'll do anything … wash cars, do chores." My brother and his friend Allan didn't want a girl partner, though, so Beanie, my nickname, became Bennie. Patrick Utley, Allan Klotz and Bennie Smith. But that was just one summer that I hid in the bushes while they went to the doors of the neighbors to deliver the "business cards" that I had typed, my male role on paper only. They certainly let me do plenty of the chores. I learned to work with boys, to translate later to work with men. I learned about which names matter and which do not. Call me Bennie or Beanie, but include me in the group. I learned about money and business and something about gender bias.

I had also been a sister to another brother who tortured me in the brotherly way. Forced to drink garlic juice laced with every spice in my mom's kitchen cabinet. Relentless teasing. Constant grilling about the miniature plastic helmets of the NFL teams of the day. "What team is this?" holding a helmet right at my nose. Miami Dolphins. I don't really care for football much today. The same brother who sent tape recordings from Vietnam and to whom I wrote about the ridiculousness of a 12-year-old's life to a GI in the jungle. He later sent me letters while I wore combat boots and slogged through mock POW camps as a cadet. Letters of encouragement, letters with hand-drawn cartoons with humorous depictions of military life, letters with stories of U.S. Army green Vietnam service to an Air Force blue cadet sister.

And I was a sister to sisters too. To an older, wiser, "mostly moved out by the time I was in junior high" sister but there for me with advice and wisdom of the eldest. And a younger, pesky, almost lookalike sister who joined me in the entertainment activities of our youth. Plays presented with a sheet hung as a backdrop from the rafters of the carport, dance routines in matching PJs, games in the yard and games on the street with the neighbors. No name changing required.

To my new life at USAFA, I brought being a daughter with me. A daughter to a mom who wore the WAVE uniform and a retired Army father strict and tough and disciplined but kind and loving and compassionate.

I brought my Spanish from life in a foreign county. I brought my experiences with exchange students from our home and from my school. I brought big-city times and small-town drama.

I brought my harvest-gold Samsonite suitcase almost empty.

* * *

The next morning, Chris and Nina wished me well and dropped me off at a nearby hotel. From there, a group of the other nominees and I climbed up onto the large silver Air Force Academy buses. The sun shone brightly off the vehicles' polished sides. Distorted views of all the faces shot back at those waving from the base of the steps.

Maybe we came by bus to initialize the group mentality, to eliminate the individual from the start. Maybe to optimize the announcing of the welcoming instructions. Maybe to avoid the emotional scenes as parents and siblings gave their final hugs and tidbits of advice. It could have been to minimize the military-civilian interaction at the drop-off point for the sake of the waiting upperclassmen who were ready and even anxious to deal with the new arrivals. These seasoned cadets were versed in the handling of the newly accepted. But they were not as accustomed to dealing with anyone outside the military rank structure. They might not have known where to place a non-military father or, even worse, the cute sister of one of the new cadets within their understanding of the rank structure. Who needed to be addressed with "Yes, sir" and who they could simply ignore. For whatever reason, instead of being driven directly to USAFA, I was added to the busload of new recruits leaving a hotel only five miles from the south gate of our new domicile and headed for the terminus: the "Bring Me Men..." ramp.

The bus ride was a short ten minutes. The Colorado Springs mayor's welcome to the Class of 1980 was on a billboard on the roadside. Again, I pressed my head against the windows, watching this new life play like a movie on the glass screen. Small, discreet signs to the Officers' Club and airfield dotted the lush green medians. Most of the passengers sat quietly. Occasionally, a voice, loud with false bravado, intruded on the silent anxiety. The large, sharp, gray mountains on the left added to the tension.

The buses pulled up into a dead end of the road. Our final arrival area looked like a loading dock, a canyon of three-story buildings rising on both sides. Stepping off the bus made me shrink in comparison to these imposing buildings. To the left was what I found out later to be the library, but at this level, there were no signs or elaborate entrances. To the right was one of the dorm buildings, Vandenburg Hall. However, at this level, no rooms were visible. The building was unmarked and unspectacular. Directly ahead was an almost two-lane concrete incline that rose one story. It was difficult to see the layout at the top. A wide marble span crossed the height of the ramp and blocked the view. It was on that span that the two-foot-high silver letters of a poem began: "Bring Me Men...."

With upper-class cadets screaming all around me, I stood in this concrete canyon and realized, for the first time, the magnitude of my decision to sign the acceptance paperwork with my father's pen. These words did not welcome me. The Air Force Academy had never welcomed women. *What would happen to me here? Would this marble portal transform me into some kind of man? Or a woman with a man's psyche?* This was bigger than my 62-inch body. This was bigger than the 156 other women who arrived at this ramp with me today to become cadets in USAFA's Class of '80. This day, this falling of this gender barrier at this all-male academy, was bigger even than this huge, imposing incline that beckoned "Bring Me Men...."

"What in the hell do you think you're doing? Get over here with your classmates or get back on the bus and get out of my academy." The screamer was in a blue uniform, but his face was redder than my hair. I grabbed my suitcase and headed toward the others but noticed his eyes were not following me. He was looking at another female arrival. His anger and attention were totally focused in her direction. I took advantage of the distraction and slid in behind a group of others not wearing uniforms. I couldn't see the front of the group over all their heads but felt safer that way and didn't want to see.

Before any of the almost 1,600 new arrivals, men or women, gained the privilege of walking up the "Bring Me Men..." ramp at the United States Air Force Academy, we had to learn the basics. That morning "Basic" became our name.

The upper-class cadets screamed "Basic" at every juncture. Later, when we all wore uniforms, I would know by the shining rank on their black felt shoulder boards whether they were juniors, "Second Class cadets or Two Degrees," or seniors, "first class or Firsties." Now their difference was more visible than two-inch silver stripes. They radiated poised and confident looks in their crisp, creased uniforms, white gloves and hats. We "Basics" meekly shuffled, sporting the common clothes of the outside world: the girl of the famous picture in her chevroned skirt and top, some young men in slacks, others in jeans, me in my calico jumper.

"Ladies and gentlemen of the class of 1980," one exceptionally pressed young cadet announced. "We are members of the classes of 1977 and 1978 who are here as cadre members for this first three-week session of your Basic Cadet Training. We will refer to this training as BCT, also suitably called BEAST. It will test whether you are worthy of being accepted into our cadet wing. Members of this BCT staff will attempt to teach you what it takes to become a cadet. You must provide the effort, the enthusiasm and the energy to make that transformation happen. Good luck."

From that first moment, I was assaulted with the foreignness of the words.

"Hey you, Basic," I heard. *I didn't know I was named that.*

"Fall in over here, you." I did not know where to fall, so simply moved with the others herding in the direction of another cadre speaker.

"Not like that, you ignorant Basic. Shoulders back and down. Stomach in. Oh, forget it, you. We'll get to that later." I hoped it was much, much later.

"All of you in this line right here," he said, gesturing to the ten new arrivals who lined up next to me. "Listen up. I am going to instruct you Basics on the only words you need to say for your entire first year. You belong to us and all we want to hear out of you are the following five responses." His voice gained volume and pitch with each consecutive word.

The concrete amplified his voice, ricocheting the commands around me. I felt more uneasy with each sound. This overzealous cadet was really getting into it. I didn't understand it then but later thought there must have been a lottery system for who got to meet the buses on that arrival day. I am sure there were many young men cadets trying to uphold the honor of the academy wanting to be the first to "set the women straight," get them "with the program." Our keen guide went on.

"You may now respond to upperclassmen, and I mean *men,* with ONLY the following five responses:

1. Yes, sir.
2. No, sir.
3. No excuse, sir.
4. Sir, may I ask a question?
5. Sir, may I make a statement?"

"There are no other acceptable words from your ugly little Basics' mouths. You may not *ever* say anything else to any of us without permission. If I ever ask you 'why?' you are to respond with, 'No excuse, sir.' If I happen to

2. Basic

give a rat's ass about any of your lame excuses, I will ask you. We will tell you all you need to know. There will be no excuses for anything else."

It seemed simple enough: five easy responses to explain away the next four years of our lives. I thought I'd try to just keep quiet and not have to worry about what I was allowed to say. Just take it all with a grain of salt, the way a Florida cadet had advised me. Try not to be noticed. It seemed like a good plan, but with only 157 women in our class of more than 1,500 and 3,000 upperclassmen waiting for a chance to teach us something, it was a plan created of delusion.

At the base of the "Bring Me Men…" ramp, on that June 28, 1976, we learned the beginning of the new language of USAFA. Some brave female decided to try it out right away.

"Sir, can I just ask a quick little something. Sir?" she said.

"Oh man, you girls really *are* dumb. A whole lot dumber than you look, and I'm telling you, you don't look like Einsteins for sure. That was *not* one of the five acceptable responses for addressing an upperclassman. You"—with a stiffened finger pointing right at my eyes—"you, little one with the ugly red hair, tell your dumb classmate what I just finished explaining."

What? I thought she did great just to have the guts to even open her mouth, let alone to want to know something else. I was frozen with fear just having them look at me. What she said hardly registered in my scrambled brain. I had no idea what she was supposed to have said.

"I don't know" came out of my throat in short squeaky bursts.

"Great, another one from the bottom of the IQ pool.

Inprocessing: Day one (28 June 1976), hour one. Arriving at the base of the ramp and learning to stand at attention, beginning to learn the "basics." From left, front row: Kathy Conley, unknown, Theresa Armbruster (courtesy Clark Special Collections, McDermott Library, U.S. Air Force Academy).

You sure didn't get in here on your looks, so maybe you're some superstar athlete. Maybe they just let any girls show up here today. I don't care what your problem is, you are not going to ruin my academy by coming in here and expecting us to lower our standards because," he continued in a falsetto voice, "it's too hard for me. What's your name, Basic?"

Catching on to "Basic" as a title for all of us, I warbled out, "Kathy Utley."

"You are not Kathy Utley, and you may never be Kathy Utley again. You are Basic Cadet Utley, and if you don't answer 'Basic Cadet Utley, sir' in the next five seconds I am going to make sure that you are never more than basic cadet for your entire Air Force career."

Lesson one for me as I shouted out "Basic Cadet Utley, sir!": Do what you're told.

It didn't matter. I didn't think I'd ever really be the same Kathy Utley again anyway.

* * *

The upperclassmen continued our reprogramming with a few of the marching basics.

"Now that you are straight on what you are allowed to say, I want you to get a few marching basics down."

"Basics, left face." Pivot on your left heel and right toe 90 degrees to the left. Pull your right heel up to the left heel sharply.

"Basics, right face." Right heel, left toe. Left heel up sharply.

"About face." Right toe back to left heel. Spin right 180 degrees on right toe and align left heel to right heel, "SHARPLY." I almost spun down in a heap.

"Left face. Right face," he shot out at us.

Shaken and confused, I couldn't think fast enough. I turned sharply but stared into the face of the line of cadets all staring at each other's backs. And at me.

"Your other right, Utley." *I should have blessed myself.*

More right faces. Left faces. About faces. A lot of just confused and shocked faces. Nothing of the way we talked or walked before would follow us up that ramp.

Eventually, we moved out. Not a straight marching column of ants. A gaggle of new recruits, including the first women ever to enter as cadets. As I passed under them, I imagined those silver letters falling in front of me, blocking my path, screaming like the upperclassmen, "We said, Bring Me *MEN*."

* * *

Reaching the top of the "Bring Me Men..." ramp, the cadet area opened up before me like a large futuristic game board. Our strict military competition would be played on this immense square of terrazzo blocks accented with intersecting marble strips all in straight lines north to south and east to west. A forbidden zone of green grass filled the center, but the rest of the view was hard, cold and made of very inorganic colors. Buildings rose on all four sides, buildings of glass and polished steel. Three of the four were squared like the center block. The fourth, the Cadet Chapel, rose with seventeen sharp spires pointed skyward. This chapel that had been considered too modernistic when revealed as part of the design by architects Skidmore, Owings and Merrill Associates. In 1955, it was likened to an "accordion lying on its side, or a line of telescoped Indian teepees," and now it was the hallmark of the area where I stood.[3] The ragged peaks of the Rampart Range rose behind the chapel, not at all inviting but at least evidence that nature, the world of living things, existed beyond this sterile zone. The entire academy design symbolized our path for the next four years: straight, rigid and severe.

The media was swarming all over the wide terrazzo expanse. This did not help our case for integration or my personal case for anonymity. The upperclassmen were in our faces, and the media cameras and microphones filled any void that upperclassmen left open. For Basic Cadet Training, there were 45 news media representatives from print, TV, radio and magazines scrambling all over the cadet area.[4] Packs of photographers and

Aerial photograph of cadet area circa 1976, USAFA, Colorado Springs, Colorado (courtesy Clark Special Collections, McDermott Library, U.S. Air Force Academy).

reporters followed our every move, our every stumble. I remember press reports that worried about our losing our femininity, stating the women were faring well but missed their bathtubs and kitchens. I never missed either. I missed my sleep.

General McCarthy later explained that in the process of preparing for our arrival, he and his team had visited the Merchant Marine Academy. Women cadets had arrived there the year before, so they were in their second coed year. "They identified for us problems that did exist. One of them was the negative effects of too much publicity. The women there were quite negative in that respect; the men were quite negative. The only thing we could do in that area, because publicity has both positive and negative impact on the institution, was to advise our cadets that that was going to happen, and hopefully minimize the impact although we got the same negative reaction from both, men and women."[5]

Commands to "keep your eyes caged, miss" scared me from looking around. I heard the cameras clicking, though. I saw the bright bursts from the flashes. It was oddly exciting. My heart was still beating wildly with nerves about the left, the right, the eyes to the front. Contrary to my own plan for anonymity, I secretly hoped the paparazzi would crowd around me and take pictures. There had been a miniature picture on the front of my hometown Florida paper, the size of a school picture that middle school girls exchange. *Ft Meade Girl to Be Air Force Academy Cadet*. I never posed for the picture. I think it was cropped from some other shot. I had never experienced anything like this. This was a spectacle.

Many of the women were photographed over and over again. The girl whose picture was taken at the base of the ramp with the suitcase in her hand became almost a poster girl for women at the academy. Others too. There were only four African American women in the group. One, Gail Benjamin, was interviewed by *Ebony* and other periodicals and was quoted in many newspapers across the country.

The presence of the media changed the environment for everyone. I'm sure the upperclassmen were self-conscious with cameras chronicling their every move. With microphones recording each spoken command, could they say what they had always said? Were they harsher for effect or quieter out of fear? Did the photographers and reporters keep us safe from extra hazing and harassment, or did the hazing arrive anyway, just out of view of the cameras and earshot of the microphones?

I never tried to be noticed, even by the photographers. Immortalized walking up the famous ramp "breaking the sex barrier" with a smile on your face would be a thrill. Some of the pictures to come later, being caught for the front page of your hometown paper with tears clearing tiny strips of skin from an agonized mud-caked face, would not.

2. Basic

* * *

I marched through the rest of the day in numbing shock. A long inprocessing schedule took little groups of ten Basics all over the cadet area. Every stop added to my confusion. At each stop, I picked up a new GI (government issue) item for my new life. At every station, it seemed I also deposited a piece of the girl that I was. *No wonder they call us Basics. That's all that is left.*

First, we turned in the suitcase that we brought.

"You won't need any of these civilian items anymore, Basics. Remove any medications, any transcripts or medical records, and leave the rest of your past life in your luggage and drop it at this corner. These items will be kept for you until you have earned the privilege to have them back."

It took six weeks to even get the hair dryer back. It did not really matter until then because one of our first stops was the barbershop.

Completely lost in the maze of halls and buildings, our group lined up outside what at first seemed like yet another office. Glancing up, seeing the red swirled barber pole, I realized where we were. The women got whatever hair we came with, no matter how much or little, cut some more, with a choice of four styles, as a show of fair treatment. The planning committee had contemplated haircuts. "The haircut problem was solved by giving the women a choice from four different hair styles only slightly shorter than Air Force standards." Prescribed haircuts from the applicable regulation were as follows:

"Hair will be clean, neatly arranged, and styles consistent with the type of duty performed. Hair may not be worn in any style which touches the top of the back of the collar. Hair will not touch the eyebrow or protrude in front below the band of the properly worn headgear, except that hair may protrude in front of the beret. 'Pixie' styles are acceptable, but short 'mannish' hairstyles are to be avoided. Braids or pony tails may not be worn. Bouffant, 'beehive,' 'natural,' and other similar hairstyles are acceptable, provided they do not exceed 2" in bulk."[6]

There was one that looked like the Dorothy Hamill bob. The others I can't remember. I had already cut my hair to a short, over-the-ears pixie style. My hair was within regulations. Only the bangs had any length left, but the shears still cut these from three inches to one inch long. The hairs went wild, freer without the weight of the extra length. I looked around at the women and could not distinguish between any of the four choices. It seemed it didn't matter which style we chose. The end results looked all the same: cropped and choppy.

Betsy Joviak remembers her hair was "close to my waist in length and then got the standard cut we all did. In hindsight, I would have gotten it cut before coming." They would have cut it anyway. Another female classmate, Gail Benjamin, remembers that she had been thinking all day about leaving. "As I walked around that day I was thinking, should I leave? Can I leave? But I have to tell you, after they cut my hair I thought, Well I've got to stay until my hair grows back anyway."[7]

I was feeling sorry for myself until I noticed the guys. Some had sported the long hairstyles of the '70s, almost to their shoulders. Now their heads were buzzed to the skin. "There's no turning back now," one male Basic whispered. They were unquestionably more branded than the women after this stop. They looked like babies, their eyes suddenly seeming bigger without the frame of their hair. They knew the hair was gone, but they could not see their bald heads. Every one of them reached up to confirm his predicament and rubbed his stubbly dome. As we marched out of the barbershop, we stepped over the men's and women's hair all mixed in multicolored tufts on the floor. We all left behind something else of our old lives on our way to becoming cadets.

During our assembly-line inprocessing, we received a little blue book. *Contrails,* the bible of the Basics, the living word of USAFA, was the most sacred possession of our new life. Worn, spine-broken, pages slipping from the binding, mine still occupies a place of honor on my desk. A small, 3 × 5 inch, blue, hardcover, 187-page book, it held the answers to most of the upperclassmen's questions. It meant salvation and partial escape if you could quote a passage, like a Protestant quoting scripture, when prompted by a passing questioner. It might even save you just by appearing open in your hands, in front of studious eyes. Holding it this way made you appear intent to learn, already disciplined. Open, it erased any look of vulnerability or some of the availability for haranguing.

 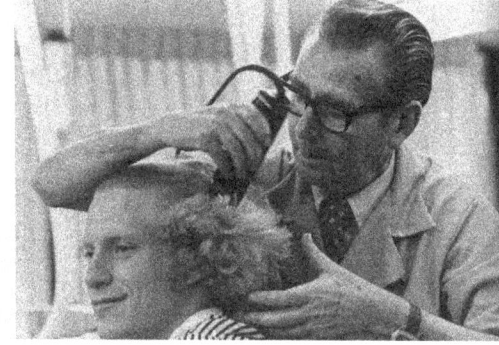

Barber/beauty shop stop. Women: pick a style, although they all end up looking the same. Men: take it all off. From left: Lenora Wong, unknown, unknown (courtesy Cadet Media, U.S. Air Force Academy).

2. Basic

We had to always have *Contrails* on our person. But even after all the hours, even years that the commandant's committee working on women's integration (no, admittance) spent on choosing the right uniform, our uniforms, a skirt in lieu of trousers, had no pockets. I held the stiff book in my skirt waistband with my stomach. This first day, we were told to start memorizing each and every page. Every available second, waiting in any building, lining up before seminars, and waiting to use or for others to use the latrine, upperclassmen barked out reminders. "Basics, pull out your *Contrails* and continue learning your knowledge."

I opened it to one of the first pages. I held it in front of my eyes as a shield. At first, my stiff stare saw only fuzzy gray lines and not words. My mind, scrambled with all the confusion of this new life, couldn't concentrate on reading. A passing upperclassman, yelling at a "gazing Basic" in front of me, refocused my mind and eyes. A picture of the academy superintendent on the left-hand page reminded me of a kind grandfather. He had become the seventh superintendent on August 1, 1974, less than two years ago, inheriting most of the preplanning and much of the preparation for the arrival of female cadets. I later heard that his entire tenure at USAFA would be remembered as leading through the acceptance of women. I read his message:

Superintendent's welcome message to the class of 1980
Welcome to the United States Air Force Academy. The staff, faculty and Cadet wing join me in congratulating you on your appointment to the class of 1980. You have gone through a rigorous selection process and have demonstrated that you have the leadership, academic and physical ability required to succeed as a cadet. The next four years will be the most demanding of your life. You will be continually challenged in the classroom, on the athletic field and in leadership positions within the cadet wing. Each of these experiences is designed to prepare you to meet the responsibilities you will encounter as an officer in the United States Air Force.
…The class of 1980 is the first

My *Contrails*, our USAFA bible. Blue was our class color. Very worn, mostly from use, some from age (courtesy Megan Q. Daniels).

coeducational class in Academy history. Your progress will be closely followed by the entire nation, and your success will reflect on the entire Air Force.

In its short, 22-year history, the Academy has earned a reputation as a distinguished military and educational institution. Academy graduates have established an impressive record of service to the nation, both in war and in peace. You are now a part of that tradition. You have the potential to make your class the best in the history of the Academy—the challenge is yours.
James R. Allen, Lt General, USAF/Superintendent[8]

I believed these words. I had just begun to feel the sting of the challenge he mentioned. This kind-looking superintendent welcomed me, even if the letters on the ramp didn't.

* * *

"Put your *Contrails* away, Basic. We are moving out," barked an angry cadet.

At the tailor shop, I was meticulously measured. I spent the rest of the day wear-

Female cadets in Basic Training studying knowledge from *Contrails*. Linda Garcia is in the foreground (courtesy Clark Special Collections, McDermott Library, U.S. Air Force Academy).

ing around my neck a white, 5 × 7-inch placard announcing: Height: 62"; Weight: 106; Head circumference: 21½"; Bust: 32". Cameramen recorded the process. The days of hiding some of those numbers were gone.

Into a green laundry bag we added the "Shape Me Sweetly" Maidenform bras and the "Miss Universe Beauty Pageant" granny underwear that we were issued.[9] In a very detailed effort to have every new Basic receive totally issued items, the academy had very purposefully made the decision to issue bras and not have the women bring their own from home. An academy representative was lingering in the lingerie department of one of the stores of the local mall. As the tailor shop staff measured the women, someone telephoned the measurements in to that airman waiting near the racks of bras. He then bought eight of the correct size of the white cotton bra. In the subsequent years, new female Basics brought their own bras to meet a standard requirement but purchased and packed from home. As General McCarthy, head of the planning committee, explained, "I felt that first year as idiotic as that sounds, that we were being consistent, and we should bend over backwards in those small things to be consistent."[10]

I didn't appreciate at the time but do ever so much now how fairness and planning went into the perception of equality that the male cadets would gain even if they never saw us in those ugly white cotton bras. Once, though, during that first year there was a "panty raid," classmate Lisa Lambert remembered. "The guys on the spirit mission knew exactly where to locate our bras and panties because the regulation book showed them!" The bras were then draped over busts of Orville and Wilbur Wright in the Air Gardens for all to see. "The worst part," Lisa remembered, "was that the cadets who returned our unmentionables knew exactly who should receive the size 32B bra because names were stamped in the left cup."[11] Utley, K.M., was stamped in the left cup of mine. After the first sweaty time I wore the bra, I pulled it off and noticed a dark bruise under my breast. Looking in the mirror, I saw it was not a bruise but an identifying stamp on me. Utley, K. M., backward on my skin but correct in the mirror. If I got too rattled by all the training, I could just check my breast to remember my name.

In August 1975, Headquarters Air Force had approved designs for the women's uniforms. The original men's uniforms, after a brief consideration of designs including lightning bolts and clouds, were designed in a classic style and in cadet blue by the famous Hollywood producer Cecil B. DeMille.[12] The not-so-famous head of the tailor shop, Mr. Peter Iannucelli, designed ours. There are some references that implied that the design was kept in-house to keep the press from finding out that such advanced plans were in place for accepting women.[13] But Mr. Iannucelli, those berets that looked exactly like mushroom caps were terrible.

For so many of us, the issue items just didn't fit. The women of 1980 came in all shapes and sizes, which understandably presented some challenges in uniform issue. Several women of my class have told me tales of ill-fitting uniforms even though we spent much time on that first day at the tailor shop. Lisa Lambert was late to the dinner table the first evening of Basic because she had to go back to the tailor shop in the evening to retrieve the issue skirt that needed hemming. That left her upper-class table companions displeased to have a Basic, a female Basic, late on the first night. Betsy Joviak remembers "wearing a skirt that was stapled for the hem length for the swearing in later that day. They just had too many alterations to get all done in the tailor shop." Her issue stockings only reached her knees, while her shorter roommate got the same size stockings which wrinkled all the way down her legs. Betsy tells her other uniform issue story to current cadets to explain some of the travails of our first year. "In my case I was 6 feet tall and wore a size 10 shoe. The process of uniform issue sent us through lines where the staff handed us the allocated number of uniform items in each category. Some things were in sizes, and some were a single size. The issue of socks (white athletic socks for our PE uniform and black socks to be worn with pants and black low quarter shoes) was in a single size," Betsy shared. The socks she was issued were too small, but the staff insisted they were "women's socks" and she could not be issued the larger "men's socks" even though they were within arm's reach. The first time she came out into the hall dressed for PE, an upperclassman challenged her on her socks, asking her loudly, "Miss Joviak, don't you understand the athletic uniform? Where are your socks?" After she ran through the allowed five responses, which didn't include a direct answer, she was finally allowed to explain that she did have socks on, but they were hidden inside the shoes. He had her bring out all her "women's socks" which were about half the size she needed. "This upperclassman was a lifesaver.... He changed them out for ones that would fit."[14]

When we'd pull that little blue bible out from the skirt waistbands, its skin side was sometimes sweaty. The skin itself had USAFA seal, eagle and all, pressed into it like an embossment. Wouldn't we have loved to have a pocket?

Thankfully, the idea of uniforms of white culottes resembling an ensemble for a female Dr. Livingstone from a trek out of Africa was rejected. The option with white go-go boots like the girls from James Bond's *Diamonds Are Forever* was also eliminated. The Air Force rejected the go-go boots as fads, with fads never appropriate for military uniforms.[15] General Beck, the commandant, the commander of our military training, during our first year later said, "If we erred at all, we erred on the conservative side. For example, the uniform our cheerleaders wore were very, very conservative as we look back now, as compared to what they could have been at the time. Nevertheless, uniforms were among the least of our concerns or problems."[16]

The civilian tailors fitted me for blues, fatigues and dress uniforms. The fabric of them all felt odd against my body. The collars choked me, and the stiffness made me feel trapped. But compared to my lifelong clearance-rack experience, the attention to each wrinkle and fold of the uniforms was unusually flattering. I stood tall and enjoyed as the tailors professionally tucked and pinned my uniforms for a regulation perfect fit.

Reveling in this attention, I heard another female voice asking, "Can't the slit in this skirt come up a little higher?"

2. Basic

Her accent was definitely from New York.

"No," the tailor quietly replied. "The regulation requires five inches below the knee."

"I'm used to above the knee," she responded.

Wow, this girl's got spunk. She didn't seem defiant, just matter-of-fact about her personal preferences. The name on her white placard read *Benjamin, G.F.* Height 63¾". Bust 33". Trying on fatigues, the green utility uniform that I knew before only from GI Joe, brought more comments from this G.F. Benjamin.

"I think this waistline needs some tucks. I don't like my clothes baggy."

Uniforms were not going to make us all the same.

Somewhere, during the day, a sergeant at the armory issued each of us an M-1 rifle. It was an almost ten-pound weapon, made heavier by a barrel filled with lead, added as lore had it to keep cadets from actually shooting up the area after ammo was stolen. Funny that lead made the rifles unusable, demilitarized, and we took them to become militarized. A sling of green webbing snapped from its barrel to its stock. Within 15 minutes, I ached to unsnap the strap and drag the fiend on the terrazzo behind me. My arm cramped from carrying it. Within weeks, I would almost caress this weapon. The military drill rifle manual would have me snapping it around to 16 different positions. I learned to strip it, clean it, march and run with it. That first day, though, we only needed to carry it to our rooms.

We picked up bags of clean linens to use later that day to make a perfectly measured, quarter-bouncing tight military bed. We stepped through an assembly line at the dispensary getting inoculations in each arm at the same time. Again, cameras recorded the grimaces. Needles drew blood into vials for testing. We exited the clinic with a little less outside world in our veins.

Laden with all the gear and confused by all the new nomenclature and rules of conduct, no longer sure who I was, I half-marched, half-stumbled toward the dorms.

3

Invader

> "Fall seven times and stand up eight."
> —Japanese proverb

"For the duration of the six weeks of Basic Training, you will be assigned to a summer squadron. Assignments of the letter A, Alpha, will be joining the Aggressor Squadron," upperclassmen shouted. "Check your paperwork, Basics. All those with a B, Bravo, designator will be members of Bearcat Squadron," and so it went through Cobra, Demon, Executioners, Falcon, Guts, Hellcats, Invaders and Jaguars.

An "I" on my papers prompted me to move toward the Invaders group. There were about 100 men looking as nervous as I felt, moving toward the blue-and-gray guidon marked with an "I." This would be my band of brothers. "Female Basics will be billeted together, separately" yelled out by one of the cadre diverted me to the women standing apart, on a different side. "The Air Training Officers (ATOs) will act as upperclassmen for the women in the women's area. You will still belong to and be responsible for all aspects of life in your assigned squadron," he announced with what seemed like a question in his tone.

I didn't care at that moment who was going to do the training. I hadn't yet come to understand, appreciate and accept the acronym ATO. I didn't spurn it yet either. I didn't know what difference it would make. Yet. It sounded like we would still be included with the men. What could it hurt? At least these upperclassmen who had been so tough today were marching off in another direction with our male classmates. We were left facing some women, a little older and in blue but not in cadet uniforms.

"Good afternoon, ladies. We are Air Training Officers here to supervise your training. All you women will be billeted in a separate area on the sixth floor of Vandenburg Hall. We will ensure all the regulations and requirements are fulfilled and maintained in that women's area. We are very proud to be part of this year. The academy is expecting great things of you all. We are too. Your training will be demanding and rigorous...."

3. Invader

I just quit listening. My feet were sore from the crazy heeled shoes I had chosen to look sophisticated. My arm throbbed from holding this ridiculous rifle, and my back was achy from all the standing up straight. I didn't care at all about tradition or pride or any of those sentiments right then. I just wanted to sit and relax for a minute.

We got a minute but not much more. We tromped up to the sixth floor with all the necessities of our new life. Florida's sea-level moist air was not this dry, over-mile-high (7,258 feet) atmosphere. My lungs burned from the climb. My thighs quivered, struggling with the last few steps. When no ATOs were looking, I pushed my free hand down on my thigh to force the leg to straighten and lift me up eight more inches. Finding our names on the doors, we split up into our rooms. Not exactly our own. We had roommates. Although mine was not there yet, her name was definitely by the door. "Walker, P.C." in white letters etched into the blue plastic strip by the door. Utley, K.M. That was me. Who was this Walker, P.C.? Patty, Pam? I had no more time to think it over as an ATO came in.

"You have work to do, miss. You better get moving. Get into your combination Delta, blue uniform skirt and light blue blouse. Fold your civilian clothes and put them in the overhead bin. You won't need those for a while. Get a blue plastic name tag and pin it to the shirt, centered on the left side. Then get started putting all this new issue away. Check the regulation book for instructions. Be in the hall ready at 1630."

"Yes, sir," I said now, already as habit, but her look of annoyance was correction enough without words. "I mean, yes, ma'am."

I pulled the door shut. It was my first moment alone that grueling day. *My God*, I prayed. *What have I gotten myself into? I didn't think it was going to be so harsh. I can't go at it like this all day and not break apart.*

But I didn't let my thoughts carry me too far under. I was actually too tired to collapse. I lay down on the bare mattress farthest from the door and just focused on the ceiling. 1630 minus twelve, that's 4:30 p.m. I grew up with a black key-wound 24-hour navy clock in the house, but I still had to do the mental math. *They'll be back too soon.*

The uniform was simple and not unlike those of some of the Catholic schools I had attended. No hair left to comb. I never was really prissy about having the right "do" or the right makeup look. Those things were gone and not much missed. I cracked the door to see other women in stockings and black low quarter shoes, so opted for the same, mistake or correct. *Better to be in with the majority*, I thought. *Less notice that way.*

I looked around the room without consciously realizing that no women had ever lived here before. I got out the only book on the gray painted bookshelf hanging on the wall above one of the desks to find the regulations. It was in a bulky, also gray, three-ring binder, heavy like the many rules each added more weight. *Weight to the book and weight on me.* More rules here than in *Robert's Rules of Order*. The proper procedure for making a bed.... Cadet procedures for addressing upperclassmen.... Here it was.... Cadet Rooms. Preparing for inspection ... aarrrgghh.... No, wait.... Proper arrangement of cadet uniform items. Air Force Cadet Regulation 50–8. Clothes on hangers all facing center. Hangers should be placed two inches apart. T-shirts folded to a 6 × 9-inch rectangle, with USAFA letters along the top fold.[1]

The arrangements kept me occupied, but the aloneness began to erode my initial composure. I heard muffled talking and laughter through the walls. I heard other women, brave enough to venture into the halls, greeting the ATOs, "Good afternoon, ma'am," then the fading of the clicking shoes. I wanted this quiet but now was afraid that I was going to have to go it alone.

As 1630 approached, I sat ready on the gray wool blanket covering my bed. *Where is this Walker, P.C.? What could have happened to her and how could she leave me here alone?* My nerves fired shocks into my already confused brain cells. *What if she doesn't come at all?* I paced and slid my dry palms against each other making friction noise. I needed someone to talk to. This isolation was rubbing at me worse than the new shoes. The doorknob moved, and I jumped up for fear of being caught doing nothing.

A smiling bright-eyed girl came bouncing in. She dropped all her bags of cadet equipment right in the doorway without even closing the door to the hall and the hostility. "Hi," she almost sang in her happy lilting voice. "I'm Peggy. I guess we're going to be roommates."

"You are late," I said. "You'd better hurry up."

I pounced right on her with the first words out of my mouth. In fact, of all the women and men she met, she only remembers meeting me. Why does memory work that way? Why couldn't she remember something funny that I said or not remember our first meeting at all? She was relaxed, and I was wound tight, ready to snap. I transferred all my anxiety to my language and smothered her smile. But not for long. She shrugged, dropped her bags and threw herself on the bed while I popped up to close the door.

She looked annoyed but not angry. She shrugged off the comment and went on with introducing herself. I hope she knew I was scared and immature but not malicious. Even those first hours, I was trying to be so cognizant of the rules, the rules, always following the new rules, that I left little time for fun. As it turned out, Peggy would be the perfect counterbalance.

* * *

3. Invader

It was Peggy's dream to fly off into space. She did a research report in the eighth grade that reviewed the lives and careers of many NASA astronauts. She copied biographies of all the astronauts and laid the papers out side by side on the floor to do some analysis. She was methodical and logical about the process. She identified that the majority of the astronauts in her sample were graduates of the Air Force Academy. *Okay*, she thought. *I'll just go there and start my astronaut career in the right place.*

In her junior year in high school, she wrote to the academy asking for application information. She was oblivious to the fact that there were no women, and had never been any, accepted at the academy. Her response from the academy was a form letter advising, Miss Walker, "you are not eligible to apply to the United States Air Force Academy: *the wrong gender.*" Disappointed and confused, she found the dictionary to look up that gender word so she could know just what it was about her that caused the disqualification. It was something about the sex of a person.

So, she pondered, *it was something about sex*. She went to her mother for clarification. "You applied to an all-boys school?" her shocked mother asked. "I guess so," Peggy confusedly answered.

She changed tack. She started paying more attention to the college brochures that came in the mail and from the high school guidance counselor. She had no money for college, though. Her father set aside only enough money for her brother to go. After her father's pronouncement, her mother found a job to save some for Peggy's education. Just after the passage of law number 94–106 that President Ford signed on October 7, 1975, a high school instructor, like my health teacher, let her know the doors of the academies were now open to women. She had better hurry.

Peggy's family was not overly enthusiastic about the academy as her training ground. A neighbor friend tried hard to convince her parents not to let her go. His brief experience at USAFA had been grueling and disheartening. She remained undaunted and even arranged a visit to USAFA. During her senior year in high school, she flew to the academy with her mother for a tour, including a meal with the cadets on the main level of the dining hall, Mitchell Hall. As she sat with her escort, she watched the ATOs being marched in and expressed her confusion. "Hey, I thought there were no women here yet." *Maybe gender meant something else.* "Those are just the ATOs," the escort grumbled. "They don't belong here."

* * *

There were 15 female cadets, among the 100 Basics in Invaders Squadron. Eight of us were in A-flight, slated to go into 9th Squadron once the academic year began. Joyce Cain, June Van Horn, Gail Benjamin, Debbie Wilcock, Donna Smart, Marianne Owens, Peggy Walker and me.

I met Peggy Walker and jumped on her with my first words. We were roommates. Gail Benjamin was the New Yorker I had seen in the tailor shop. She lived next door with Marianne Owens who felt she "had two great supportive roommates with a war going on outside our door."

Marianne's dad was in the Air Force. She grew up all over. "At some point during my childhood, we drove past the academy, I looked at it and my dad said, 'Someday, Skip (my brother), maybe you'll want to go there.'" Sometime early in 1975, some Navy admiral suggested that she apply to both USAFA and USNA since they were opening up to women. Her dad was concerned about the opportunities she would have. He thought that "women were not given jobs commensurate with their talents." He believed, too, that cadets would resent the women. Her mother wanted her to go to college and have a regular sorority life. Marianne attended a meeting for prospective cadets at Bolling Air Force Base before her arrival and heard current cadets saying, "We don't want the women. They are never going to survive, and if they make it in, we'll make it the hardest year of their lives." Marianne's mental response: *"For anybody with a spine, it's going to make them say, What? Are you kidding? Of course, I'm going to go. Of course, I'll do well."*[2]

* * *

Gail stepped into our room and introduced herself shrugging her shoulders and rolling her head about all the hoopla in the halls. "What's wrong with all of them?" she asked. She was our strength. An older, wiser head when all about her were losing theirs. She had attended Vassar for two years. Her funds dried up, and she had to leave school to go to work. She was working a few jobs to save enough to go back. Her brother, a 1975 graduate of USAFA, remained at the academy after graduation in a position with the Minority Affairs Division. When the legislation passed to allow women into the academies, women were classified as minorities—a no-brainer as there were none at the academy then and there would be only 10 percent the first year. Lieutenant Benjamin recruited his own sister. "It only took a couple of words to convince me," she related. He explained, "They will pay for your education," and she was hooked.

She was unflappable. "Every time an upperclassman yelled in my face, asked me to identify an aircraft or told me how they hated women being at the academy, I would swallow my smile. I gladly did what they commanded and took all they had to offer. I'd just think, *Someone else is paying for all of this. Someone else does the laundry and I don't have to worry about where I can afford to eat.*"[3]

She later described the beginnings of our time as "not welcoming, with friction" and "people did not appreciate my human packaging, the color of my skin or the fact that I was a female. I learned to not focus on that."[4] What a warrior and a challenger too. She questioned things from the perspective of a little more experience than the rest of us. She would actually ask, "Why do we have to do that?"

During our first day of inprocessing, when upper-class cadets were yelling at all of us to "fall in" and "move faster" and "cage our eyes," Gail was confused. She handed her transcripts to an upperclassman behind a table along the assembly line. She announced, "I am not really with these people. I am a transfer student." He laughed and told her to get over with everyone else. She tried again asking to see an officer in charge and telling him that there was some mistake with the paperwork. She was an incoming junior. No mistake, he assured her again with a grin. It is a four-year program for all. No one enters the academy except through the basement door, all are freshmen or "Doolies," for the required difficult year regardless of experience or schooling. She was the only cadet I ever knew to have had the leisure time in her upper-class years to take sculpture and painting classes. She surely had most of her core requirements completed by her junior year, though probably not much earlier because we were all required to take engineering courses regardless of field of study. Her brother didn't tell her the whole story. She didn't have to pay, but there would be a cost.

* * *

It seemed like it was the middle of the night. The sun was definitely still sleeping. At "o-dark-thirty," in our new vernacular, we heard insistent beating on the door. "Wake up, you rotten Basics, and get yourselves out in this hall, full fatigues, combat boots, under arms in five minutes or we'll be in there dragging you out."

"My God," Peggy mumbled, "it can't be morning yet. They just left us screaming out there a few minutes ago. Everything hurts and I ache from not sleeping. Please, Kathy, cover for me. Let me sleep just ten more minutes."

"You know we can't, Peggy. Come on, get going," I reluctantly said. I wished we could hide here, behind the door, too.

We heard the beating moving down the hall with upperclassmen waking the other Basics in the women's area. Our first day was a few weeks ago. Although we had learned most of the routine by now, we were in no way accustomed to it. It was brutal and there was no becoming accustomed to it. We did it through sheer will and fear of the repercussions if we did not. I learned to fight my mind, my body and my tears through most of it.

"Get going, you lousy SMACKs. I'm coming in after you in one minute, dressed or not," bellowed the upperclassmen pacing like caged animals along the outside of the door. We hadn't memorized SMACK as

"Soldier Minus Ability Courage and Knowledge" yet, but by the way they were screaming the word, we knew it didn't mean anything endearing. But the upperclassmen wouldn't bust right in the room. They waited to throw the door open until the moment we announced that we were both minimally clothed.

We did have that advantage over the men of our class. They faced the upperclassmen with the first eye-opening awakening. For the Doolie men, the first screams and the door openings took place simultaneously. With the inclusion of the first 157 women, those rules all changed. The upperclassmen could not open our doors without knocking or enter our rooms without asking if we were dressed. Still, it was just a matter of minutes before the ATOs arrived on the scene to burst in and demand that we hurry. The ATOs did not run interference or protect us. They stood in our rooms and kept vigil to ensure no talking or stalling kept us from jumping into our uniforms. They threw the door wide when our shirts were barely on. Then the male upperclassmen could finish shouting their morning assaults within inches of our faces instead of through the wooden door.

To try to avoid the ATOs altogether, we hurried to pull off the issue light blue pajamas and pulled on the green fatigue pants before their arrival. Splashing water on our faces from the sink in our room, we added to the ensemble the military-issue 1940s–style white cotton bra. Over the name-stamped bra went a white T-shirt stamped with our last names right over the blue USAFA letters on the front left side. We covered the T-shirt with the long-sleeved fatigue shirt even though it was summer and the outside temps sometimes soared in the middle of the day. We didn't much mind in the a.m., though. It was cool before the sun came up. We finished dressing with the fatigue pants tucked up into the rubber bungee "trouser blousers" that held and bloused the bottom of the pants at the top of our black leather combat boots. We were ready in just a matter of minutes.

"Yes, sir. Yes, sir. We are almost ready," we piped back while they pounded on the door and rattled the knob with their angry voices rising in decibels.

"Open this door now. Get going, ladies. You're late and you are pimping over your classmates."

Pimping, also part of our new vocabulary, didn't mean that we were offering our sexy bodies for sale on the street corners at night. No, that choice phrase meant that something you were doing or not doing was going to cause some harm (usually in the form of extra yelling) to your classmates. In this case, supposedly, by being late only five and a half minutes after we were awakened, we were pimping over the rest of the entire cadet wing. Of course, they screamed that every morning, so it lost lots of its potency as a threat. No matter when you got up or how quickly you jumped into your fatigues, fireman fashion, you were still late and still "pimping over your classmates." At first, I imagined all of the men of our class were already dressed in perfectly creased fatigues. Their rooms spotless, their rifles shining like the spires of the Cadet Chapel, they were waiting for Peggy and me out on the terrazzo. I could see them being made to do millions of push-ups and rifle inspections until we arrived. How did they make it outside so much earlier than we did? We had no hair dryers and no hair to use them on, no makeup and no jewelry clasps with which to fumble and no choices of outfit, as the uniform of the day was the same for the whole training session. Within three days, we were getting dressed without even a thought of what we were putting on. How did they always beat us out there, and why were we always late and always the ones doing the "pimping"?

We were tired and confused but not stupid. So after a few mornings of the bellowed accusations of slowness, arriving panting after a brisk run down three flights of stairs to terrazzo level to "form up" for the morning run only to find male cadets arriving after us, we caught on. The constant accusation of lateness was a scare tactic and not an assertion of reality.

We also learned not to rush out and leave any morning duties undone. We left the beds made to a drum-like tension. Most times, they had not technically been slept in. We slept on top of the perfectly made and measured bed with an extra sheet so all the rigmarole of making the bed to regulation perfection would not slow us down. One of us quickly smoothed the evidence of the body weight out of the scene pulling the blanket snug from below. Anyone's hands would do. One of us would stuff the used sheets under the false cover on the laundry bin. We both were responsible for the condition of the entire room, and we both added to our total demerits toward greater punishments if any demerits were "awarded" for a messy room.

"Don't forget, 'under arms,'" Peggy yelled. Not a reference to deodorant, she was reminding me that we had to carry our ten-pound M-1 rifles this morning. These we reluctantly grabbed from the rack by the door. They were heavy accessories, and we would rather walk by them on our way out.

In "The New Cadets," in *Airman* magazine, Colonel Spruill, the commander of our Basic Cadet Training, described the challenge as "42 days, 16 hours a day, every day, with no free time at all." He explained some of the concessions made in the training because of these "new" female cadets. For one, the bolt springs of some of the women's rifles had to be changed from eight pounds of pressure to five. "You could just see the frustration on a small girl's face when she couldn't get that bolt open.... They showed frustration more often than the men. It seemed like they wanted to excel very badly and got upset when they weren't able to fit their perception of what it took to measure up."⁵

I felt that frustration and the ache of the black-and-purple bruises on the heel of my left hand. I don't remember the bolt spring pressure being changed. Many times during in-ranks inspections, a cadet would walk along the line of Basics and pivot to stand in front of me, toe to toe and usually chest to face. At the command "Inspection arms," we were supposed to let go of our rifles with our left hand leaving it balanced in the right and slam our left thumb pad down along the operating rod to open the chamber for inspection. My chamber was always clean. I could strip a rifle to its parts, oil it and put humpty-dumpty M-1 back together again. However, I could rarely get the bolt open on the first attempt to prove it. "Try it again, Utley," the cadet inspector would bark, and sometimes I could and sometimes my hand just bounced back off the metal tab again. The inspectors would usually walk away after that, some in disgust and some to allow my throbbing hand a break.

Basics running the marble strips, white gloves, under arms. Upperclassmen in dark uniforms are observing and correcting. Karen Wilhelm: "my least favorite activity!" (courtesy Clark Special Collections, McDermott Library, U.S. Air Force Academy).

Peggy and I, under arms, gave each other a quick inspection, a "checking off your roommate," and were satisfied that we were in order but by no means ready to face the day. We opened the door and quickly marched out and against the wall with a chorus of "Good morning, sirs," really meaning "Okay, here we go again." The other A-flight Invader ladies were lining up there too: Gail, Marianne, June and Debbie.

"It's about time, Basics. You ladies are going to have to pick up the pace a bit or you will still be in Basic Training when the rest of your class graduates. Okay, let's get out there. Right face…."

But one morning, Gail stopped him in mid-command. Punching her fist straight out in front of her to indicate a question, she asked right away, "Sir, may I ask a question?"

"Benjamin, what is your problem?" he asked incredulously. "Aren't you listening? We've got to get going."

"Yes, sir. Sir, I still need to ask a question."

"What is it, Benjamin?"

"Sir, may I go to the latrine?" She had not yet had the opportunity, or maybe had not taken the time or risk, to walk out into the hall to the bathroom at the corner. We could have gone. It was not *not* allowed; maybe she wanted a little more anonymity behind that closed door. Knowing Gail, maybe she was just behind schedule.

"No, Benjamin. We're late. You stupid girls took so damn long getting your ugly selves ready and now you have pimped over all of your classmates who are outside ready while you female Basics are goofing off up here. I can't believe they ever thought that girls would be able to hack it here." They rarely called us women. Even in the newspaper, the headlines would read, "Girls at the Academy," while the male cadets were almost always referred to as the "Men of the Air Force Academy."

But Gail persevered. Her fist again shot forward. "Sir, may I make a statement?"

Screaming now, his face boiling with redness, the upper-class cadet lost it. "You just don't get it, Benjamin, do you? Your self-centered attitude is going to make all of us late but even knowing that you still have something you think is more important than all of your classmates. You selfish women deserve all the harassment you get and more. I, for one, cannot wait for the day that we see the last one of you heading out, down the 'Bring Me Men…' ramp and real academy training can continue. Go ahead, Benjamin. I can hardly wait to hear it now. Tell us all what is so damned important. Make your statement."

With her posture ramrod straight, her chin in against her neck, in her most respectful and professional voice she announced, "Sir, I need to go to the latrine, as I just started my menses." For some reason, we all had this proclivity to use absolute correct terminology when addressing the upperclassmen. I glanced over at her without moving my neck. She seemed perfectly composed and unembarrassed.

The upperclassman dropped his eyes to the floor to hide his confusion and discomfort. His posture broke from the stiff superior pose to one with a curved spine, a hanging head. He was shaken, seemingly in body and thought. A quiet pause lasted just a few seconds, then he excused Gail Benjamin, cool and composed, to go to the latrine. In fact, we were all ordered, as his ramrod posture returned, to "get out of my sight, Basics. All of you. You all just go to the latrine."

"Yes, sir," we answered and all trotted off to the bathroom at the end of the hall, guiding our right shoulders tight to the wall the whole way. The door still closing behind us, we collapsed into muffled laughter. June slapped high fives for a minor victory over the oppression of the upperclassmen. "What did they think?" June snickered. "We are all on our periods at the same time? Maybe he thought he could get the whole group covered and not have to deal with the issue again for a month."

"I haven't had my period since I got here. My body is too nervous to add another function. It is either marching or menses," I said. Four of the women agreed.

We enjoyed the small reprieve and a chance to let the shoulders sag and the chins hang out for a minute while Gail took care of her "menses" problem.

I felt resentment underneath the minor victory, though. We were only allowed to be women when it explained our weakness or inability to perform. The men would be even more convinced that we didn't belong. Who was I mad at? Not Gail. I admired her for being tough enough to face them with an issue that was clearly female. I also wished my body would not rebel, would be strong like hers. Not that I wanted my period. Missing that in these circumstances would be insane. But I wanted to run and still have energy to chant out loud, to have breath to encourage others like Gail did. I was mad at USAFA for expecting us not to be female. I was angry with the cadre members seething beyond the door for blaming us for who we were. I wanted to be a part of it all, but I was still going to be a woman. How was that going to work?

We couldn't dawdle in the safe haven of the bathroom long. We knew we *would* eventually be the last ones and our classmates *would* suffer in the waiting. We were part of a team. We understood that, and we played willingly. We fought to be part of the entire cadet wing and tried not to lose too many battles in the effort to win the whole war. We steeled ourselves, took a last deep breath and erased any semblances of remaining smiles. In our most stiff military posture, we returned to the wall outside.

3. Invader

We marched to the stairwell, down the stairs to the terrazzo level and then jogged to our place in our squadron among the men in Invaders and all the others in A–H and J.

* * *

The officer responsible for the plans for those women's latrines was the father of another cadet, Derek Hess, who lined up with us behind one of those other Basic Cadet squadron letters. The plans for the latrines were part of the bigger plan to integrate women into the Air Force Academy.

Colonel Robert (Bob) Hess, Derek's father, was one of the original ATOs who trained and mentored the first class of cadets back in 1955. Colonel Hess then worked for USAFA's superintendents from 1972 until 1977, and directed the planning for the admission of women to USAFA. "I was the only former ATO assigned to the USAFA at that time—which basically made it easy for me to write the plan."[6]

Learning of the detail and thought that went into the planning for our arrival contrasted with what I saw with my own eyes and felt with my feet and heart: every flaw in the actual execution of the plan. I never appreciated all that was thought through and discussed and analyzed to yield all the many things that went right.

The planning actually started in 1972 just after Colonel Hess arrived at the academy. General Clark, the academy superintendent from 1970 to 1974, came back from a meeting at the Pentagon announcing that there was a lot of discussion about admitting women. Hess worked for the director of operations and said the superintendent "called us into the office and said we better start a plan."[7]

The operational name of the plan was Operations Plan Number 36–72 Integration of Females into the Cadet Wing. The printed plan had the Air Force seal on the front along with a sketch of the cadet area with the mountains framing the buildings, but because of the pink cover, it was known as the Pink Plan.

All the plans were color coded according to Colonel Hess. "We wrote contingency plans that were going to be acted upon soon and they had yellow covers. If it was a longer-range plan, they had green covers. If they were emergency action plans the cover was red," he explained. "We were not sure what we should have for the plan to admit the women, and one officer who had taken my place in long range plans said, 'since they're women why don't we use pink for the long-range cover?'" Hess asked if they even had pink paper down in the printing plant on site in Harmon Hall. The officer returned about an hour later to report that there was no pink paper, but they wouldn't have any problem getting some. Hess had them order enough to print not only the cover for academy needs but enough for worldwide distribution.

The plan was written and rewritten throughout the years. Each iteration got a new plan number but still a pink cover. I did once see a cover with the plan name, a plan number and an image of a female in a bunny-shaped swim ring all on pink paper. She was wearing a bikini with the top half falling off. I believe it was a joke cover for an upcoming "pink" upheaval.

The plan had three phases. "Phase I. Preparation" in 1972 (no physical changes). "Phase II. Transition" after the law was signed on October 7, 1975. The final phase, "Phase III. Execution," was further explained as "we anticipate modifications of programs during this phase to continue to be made as experience is gained in the training of women cadets in a service Academy environment."[8]

This last phase was our phase. The guinea pig phase where we put the plan into action and affirmed its expectations in some arenas and rejected some others. The best laid plans....

There were many inconsistencies in execution, despite all the planning for our arrival. The planning committee, for example, after one of their visits to other military schools already training women, saw the

Plans for integrated training with "no skirts with rifles" didn't always actually pan out (courtesy Clark Special Collections, McDermott Library, U.S. Air Force Academy).

women in skirts marching with rifles. The inconsistency of a skirt and a rifle didn't sync for the head of those visits, so they adopted a policy of USAFA women cadets wearing pants when carrying a rifle.[9] Those thoughts and forecasting seemed invisible to me at times. They were necessary and appropriate, but when we went to class in skirts, then just the women had to run back to our rooms to change to pants for drill practice which included rifles, I thought no one had planned for us at all.

There is a picture used often to portray those early days of our time at USAFA. The picture shows a group of female Basics marching up from the field house after some physical fitness testing. The women are marching in a flight on their own. They are being directed by one of the ATOs dressed in her PE gear, white shorts and white shirt. The female Basics are in their PE gear too, blue shorts and white USAFA T-shirts. In the periphery of the photo are flights of our male classmates marching up from the same testing. They are wearing the same PE uniform as the women. But the added uniform item for the women is apparent: raincoats. The courts look wet and shiny, so evidently it had rained, but it looks so out of sync for the women to have raincoats on when no one else did. Some of those discrepancies impeded our assimilation and made us stand out as the women who needed to wear raincoats on a wet day while the men could get soaked.

Consequently, part three of the planning, the execution phase, was our challenge. The "we can't wear skirts all the time" phase. They mark us as different. The "we can't be the only ones wearing raincoats" phase. They mark us as pampered. The "we need to live among the men to dispel penthouse rumors of luxury living and lenient treatment" phase. The "we can be intercollegiate athletes," "we can handle Basic Cadet Training," "we can bond with the men of our class, and we can succeed" phase.

Hess asserts that they were updating the plan all along. They sent a copy to West Point and Annapolis. "Later the Army and Navy decided they shouldn't have a plan; they should fight the admission of women to the academies," Hess says. Once, later in the planning process, Hess met with representatives of both West Point and Annapolis. "I briefed our staff and also West Point and Annapolis, and West Point also brought and briefed their plan." The admiral from Annapolis put up a single slide of two Navy men at a table, in the background a porthole with ships sailing by. The caption said, "If God wanted women to be come to the Naval Academy, he would have made them men," Hess recounts.

Hess remembers that the air staff in Washington made one change to our plan. "We wrote the initial plan as the integration of females to the cadet wing. Male and females are species, so we needed to refer instead to men and women. All the current documents and publications were written as male cadets, so all the regulations and our plan had to be changed

Raincoats for the women: a perceived sign that we were being pampered. Front, from left: Chris Reasoner, Andrea Bopp (courtesy of Clark Special Collections, McDermott Library, U.S. Air Force Academy).

to men cadets and to women cadets." The plan title changed too. "Initially it was *The Integration of Females into the Cadet Wing* and we changed it to Operations Plan Number 76–75, *The Admission of Women into the Cadet Wing*. The plan was modified up until the day we published it." In fact, the copy of the plan I have has a pink cover dated October 7, 1975, the exact date that President Ford signed the bill to allow women into service

3. Invader

academies. The printing plant was busy that very day printing with pink covers.[10]

Thought went into so many details. Cost was a factor, but certainly, factors to accommodate gender differences were paramount. Converting latrines from men's to women's was essential. The urinals in the men's latrine needed to be removed. According to the civil engineers, it was going to be costly, so they decided they should just cap off the urinals and cover the pipes with full length mirrors. The project was completed with much less cost plus the price of some full-length mirrors.[11] I appreciate their attention to detail even though it was many months before I cared to or had time to look in those mirrors. In later years, with more women cadets spread throughout the dorms, other latrines were converted by just adding a pot of artificial flowers to the bowl of the existing urinals

* * *

Back behind the Invaders guidon after our menses break, a morning run took us, all the Basics, three laps around the grassy quadrangle of the terrazzo. We ran in squadrons, in the ten giant green cubes of men and women together. An airplane in each corner framed the forbidden park in the center of the cadet area. Panting from the run, a command of "fall out" dismissed us at the doors of the cavernous dining hall, Mitchell Hall, that provided us with three square meals a day. They were square in every respect. There was no part of the meal process not accented with corners and straight lines. Basics by squadron had table assignments, but we were only allowed to walk along perimeters. We marched along the walls to an intersecting aisle and then did a body-whipping 90-degree spin to left or right and a stiff, ungazing walk along the right side of the aisle to the table. Finding the table number without moving your chin left or right and without getting caught with your eyes roaming was a feat. Classmates helped each other, quietly whispering a table number when a lost soul would pass to alert them of their current location.

When we ate, we were required to maintain a square sitting posture with spine parallel to the back of the chair, body fist-distance from the table, chin parallel to the floor. Awaiting the beginning of the meal, we hoped that the upperclassmen at the table had better things to think about than us. But they inevitably had a myriad of questions about the "knowledge" we were to have memorized from our little blue bibles, *Contrails*.

"Give me the next Air Force Day, Van Horn."

"Miss Wilcock, I need the quote from Patton."

"You, Mr. Woodlands, tell me all about the KC-135."

"Lalusis, who makes the F-4?"

They inquired and we recited facts and figures, often over the voices of our own classmates who answered other knowledge questions. A drone of words and facts filled the hall like an airplane engine revving for takeoff.

A call for "Cadet Wing.... Attention" left us unaffected; we had been at attention the whole morning already. The call did, however, divert our questioners to a stiff attention at their own places to focus on the announcements of the day. As soon as we were all released to "take your seats," the waiters came by with hot trays and platters of breakfast delights. We Doolies at the lower end of the table received the food. We eyed it and smelled it but then dutifully passed it all to the head of the table. When the upper-class cadets had served themselves, they gave us the trays and permission to eat.

The eating upperclassmen left us somewhat to our own nourishment as long as we made no noise, moved very little and maintained the square posture. Every bite made its way to our mouths after a stiff straight-line trip to the lips with our heads as far from the plate as our necks and spines would allow. After 15 short minutes, a class color light, ours blue, mounted on the wall blinked on. When that semaphore signaled "time's up," the upperclassmen summarily dismissed us. By opposite return route, we made our way to the doors. Passing through the Mitchell Hall threshold to the outside, we formed up in 10-person elements or 30-person flights and jogged to our next event.

Coincidentally, the academy instituted a new training philosophy of positive motivation the same summer we arrived. Colonel Musser, the deputy commandant, explained that "in the past one of our objectives had been to test a cadet's commitment during BCT. However, we discovered that at an average age of 18 years, 4 months, most Cadets have not yet committed themselves to a military career." To help the cadets to develop their commitment and motivation for a life in the military, the academy was moving away from "squat-thrust" leadership to a leadership style of positive motivation, a philosophy of "realism and relevance" that cadets would be able to use in their careers in the military.[12] It makes sense. I cannot imagine that as brand-new second lieutenants, cadets graduating from USAFA would be able to motivate any airmen to enhance their performance by asking them to do squat thrusts or by yelling in their faces. General Allen, the superintendent upon our arrival, assured that "of course, we did implement that concept [positive motivation] the following summer [summer of '76], coincidentally with the admission of women. As I have pointed out before this, it was purely a coincidence. There was no cause-and-effect relationship whatsoever."[13] I never knew the before so cannot comment on the after, but the training was plenty harsh with much negative motivation. Of course, this purported easing of training and softening of philosophy, however appropriate and relevant, was blamed on the influx of women, which made some react more severely.

During the first three weeks of Basic Cadet Training, we spent our days in the cadet area of the academy. We blackened-in reams of bubbled answer sheets with a number two pencil for tests in math, language and science for placement once the academic year began. It is a wonder any of us performed above an eighth-grade level, even though many were the brightest in their high school classes. We were tired, disoriented and constantly reminded of how little we knew, at least in the military sense.

Upperclassmen were absent during the testing, so we relaxed. During breaks, we talked to each other, men and women, and swapped stories of BEAST so far. The stories were basically the same, tiredness and hunger the prevalent themes. "I had no idea it would be this tough" or, "Man, jail is looking pretty good right now." Being together and letting down our nonexistent hair helped us build the camaraderie that made us the Class of '80, regardless of gender or squadron. Like all BCT classes, we bonded because of our common goals and our common enemies. We understood each other through tears and triumphs. I guess some may have competed and mentally kept score, but I had no time or energy for that.

We spent other early days addressing all of our military deficiencies. We learned the commands for marching. These commands intended to take the gaggle that we were as we amassed in our squadron groups of 150 and refine us into a marching machine, moving together in unison. We learned to step forward always on the left foot. We learned to march with the cadence to avoid bouncing out of step with the rest. No Thoreau's "marching to your own beat" in the military. We learned to count two beats after the "Ready halt" command. I often missed this one and plowed right into the cadet in front of me, jarring all up the line like a row of dominoes. We learned left and right as if we had never known them and many times left one lonely Basic marching to the right while the rest of us marched left. "Your other left, Utley," an upperclassman would announce for the amusement of all the tourists watching nearby.

Near noon, all the Invaders ran back to the formation area that staged us for the noon-meal formation. This was the main parade of the day. The entire terrazzo and living area were off-limits to visitors, but from their perches 30 feet up on the chapel wall, with the chapel's steely sides as their backdrop, the tourists could look down and see us about our daily routines, a well-drilled and disciplined army of blue Air Force ants. We caught only glimpses of them from the corners of our eyes. "Gazing around," even if only with a small motion of the head, was a gross infraction. I had little time or interest in gazing anyway. Focus kept you anonymous and sometimes out of the limelight for some small moment.

Tourists crowded shoulder to shoulder on the wall with plenty of reporters and photographers scattered in the swarm. Some with press credentials also breached the terrazzo getting closer to take pictures. Photos showed female cadets without male cadets blocking the shot. Sometimes the photographic history made it look like women and men were cadets in equal number or, worse yet, that there were almost only women in the Class of '80. One male classmate tells of being in *Newsweek* magazine in July 1976. "My mom still saved the article," he said, "but it was about the first class with women. I was only in the picture because I was standing behind one of you." They did stand behind us, in front and mostly alongside us as we all struggled to find a place at USAFA as the Class of '80. When Keith Odegard, at 6'4" marched in the front right of my Basic squadron, Invaders, he had to be taking baby steps as we marched or ran. At 5'2", in the back left, I would never have been able to keep up if Keith and the other men sized by height to the opposite corner had not slowed the pace for us.

Peggy and I arrived in formation as required, five minutes before the final call. Timing was critical because being late was cause for a write-up and demerits. Being early was just as dangerous. Upperclassmen descended on the early arrivals for training and knowledge inquisitions. Sometimes, with the ratios of upperclassmen to Basics way out of balance, Basics' heads were like Rain Bird sprinklers flipping back and forth as

Doolies marching in squadron formation to Mitchell Hall for lunch. Women in skirts and men in pants. It was tough to blend in. Janet Libby (facing camera) (courtesy Clark Special Collections, McDermott Library, U.S. Air Force Academy).

an upperclassman positioned on each side fired questions and pointed out uniform discrepancies.

"Cadet Utley, you are a disgrace."

I thought I looked regulation sharp that day. My gig line was straight: the line of my shirt, the edge of my belt buckle, and the fly of my pants were one vertical stretch. I remembered to ask Peggy to give me a "tuck." She pulled the back panel of my shirt taut and pushed the excess fabric into tucks at the sides. I then gave her one. Before we left the room, Peggy checked me off. I checked her off and we headed out confident of my appearance. *Holy moly, what had I forgotten? Lord, please let my shoulder boards be on my shirt, on my shoulders, and right side up.* Of course, I didn't dare reach up and check. And I couldn't see them. With chin in and shoulders back and down, even a flounder would have difficulty checking on that possibility.

"You have a gross cable hanging from your shirt, Utley. I am totally disgusted by it. May I touch you?"

Interesting question. I wonder if they asked the all-male Class of '79 that? Later, Gail's brother Philip confirmed that the phrase had been the same queried permission while he was a cadet before his graduation in 1975.

A gross cable was not, as it sounds, really that gross or even a large twisted rope. My gross cable was a two-inch thread hanging from a shoulder seam, on the back of my uniform.

"Tie this cable in 150 knots, Utley, and return it to me by 1900 this evening. And get your roommate to check you off next time."

"Yes, sir," I answered seriously but wondered if this was a real task or just a test of my ability to follow orders. It was impossible to get 150 knots into this thread. If I didn't report back, would he forget or come looking for me with a more difficult assignment? These conundrums were worse than the physical trials sometimes. I often just couldn't quite figure out what I was supposed to do. That was essential academy training that I didn't understand at the time but grew to appreciate and admire. Time juggling, prioritizing and sorting through the frivolous and the essential. I licked my fingers, rolled the string into a ball and stowed it in my pocket in case I needed it later.

In formation in front of the dorms, the upperclassmen capitalized on more training time. They displayed miniature slides of airplanes in front of our eyes.

"What is this airplane, Owens?" I overheard. *Please don't ask me.* I hopelessly wished for invisibility. I hadn't even gotten to the *Contrails* pages describing all the airplanes in the inventory yet.

Marianne obviously had. She spewed the answer, machine gun style: "Sir, that is a Rockwell OV-10A, Bronco."

"Good job, Owens." He turned in my direction.

Bracing for the tirade, knowing I could barely even recognize a hot air balloon, I was saved by the "Fall in" command echoed from one cadet group commander to another, down the cadet chain of command until the command reached the squadron commander standing alone in front of the Invaders.

In the evening, we often headed straight to Arnold Hall for a motivational activity. Guest speakers told war stories. High muckety-mucks of the academy military hierarchy told us of our responsibility to uphold its high standard. What really motivated us was a time to rest a bit. Tonight, rest disguised itself in the form of the movie *Patton* with its motivational message. The message was lost on most of us as the lights dimmed and our eyelids closed simultaneously. Many slept through the entire 171 minutes as the battles raged on the screen. Some Basics feigned alert attention until the movie progressed and the upperclassmen quit checking on the bobbing heads. I was startled awake by the cheering from some audience members, probably upperclassmen still awake, when George C. Scott's gravelly Patton voice bellowed, "Now I want you to remember that no bastard ever won a war by dying for his country. He won it by making the other poor dumb bastard die for HIS."

We returned to our room late and had no time left for shoe shining or letter writing. Some still wrote after taps under their blanket by flashlight. Those were rewarded with return letters in the days ahead. By unspoken unanimous agreement, Peggy and I, room 6A21, just quietly lay atop the gray issue blankets folded to a precise 12 inches from the top of the mattress with the extra sheet retrieved from its hiding spot for cover. Without flashlights or writing paper, without little blue bibles at hand for memorizing, Peggy and I waited silently for the trumpet blare of taps to end this day of our Basic Training tenure at the United States Air Force Academy. We didn't wait long for the arrival of the dreams of home again, possibly with a little Patton jumbled in, conceding that soon would come the firecracker awakening to another comparable day.

4

Pugilist

"Ignis aurum probat, miseria fortes viros.
Fire is the test of gold; adversity, of strong men."
—Seneca

Part two of Basic Cadet Training, we spent in an outdoors camping setting, Jacks Valley. The Jacks family used to own the land; it was a 3,300-acre training area within the academy perimeter. It is in the area near the north gate of USAFA about five miles from the cadet area.

It was a slightly more relaxed environment, but there were no songs around the campfire or s'mores. It did afford us a break from the cadet area routine. After three weeks acclimating to our new wardrobe, we left the more formal blue uniforms behind and put on our green GI fatigues and our black spit-shined combat boots. These same boots, worn mostly for morning runs up until now, were the reason I could not join the rest on the march to the valley. "Sick call" was my designation that day. In cadet speak, I was "walking wounded." My boots stayed shiny in my duffle bag while my classmates marched, ran, chanted, and fouled their shines with dust on the way to our new home for the next ten days.

I watched from my dorm window as all the cadets formed up to head out to Jacks Valley. I didn't want to go with them. I hated running. Honest with myself without witnesses, I knew I was not unhappy to be missing the trek. But the tears slipped from my eyes again. I had always been oversensitive, but I was exhausted with the crying. My eyes were swollen enough without the tear-induced redness. My body was dehydrated and cracking enough from the dusty thin air. Ridiculous. Crying about something I didn't even want to do was ridiculous. Plus, I had a medical excuse. Sort of.

I had succumbed to medical scare tactics. I am not sure to this day if I could have made it. I think I should have tried. My legs had capitulated in the war against the black combat boots, the morning runs, and the hard terrazzo of the academy. The pain took me to the cadet clinic. The doctor's diagnosis was shin splints.

"You are tearing the muscle away from the bone with each step," the doctor told me. "You need to stay out of those boots for a while to let your body recuperate."

Sick call, the walking wounded, may have been the answer for my body, but it was damaging for my mind. Thus labeled, I joined a group wearing sneakers that was escorted everywhere, separately from the rest. Not participating in the body-pounding morning runs or marches didn't hurt but feeling left out after the run was over did. I had a part but was not really able to revel in the victory. I lost common involvements. This was a shared academy experience and a tough personal dilemma. When to opt out for health reasons and when to power through so you could share in the involvement, be part of the mass, be part of the class and the history?

There was the added burden of knowing that my actions reflected on the women as a group. Many in the walking wounded lines were women. According to the cadet clinic, "During basic cadet training, female cadets did incur a higher incidence of tendonitis, shin splints and stress fractures than did the men, however, the majority of those problems were minor." At least they also included, "Experience revealed that women are physically capable of handling the rigors of cadet life."[1] The boots were stiffer than the men's and caused many to have those tendon problems with their heels. Ill-fitting socks produced blisters and proportionally heavier rifles added to muscle strain. Yet these "excuses" were not allowed. Women were judged as weaker and unable. I did not want to be a contributor to that false conclusion.

Arriving by bus not long after the healthy marching classmates, I did not escape raising the tents for our stay in the dust and dirt. Bigger than our dorm rooms, one green canvas tent held all 15 women of Invaders. We placed poles around an oversized rectangular wooden pallet and pulled and stretched the tent from the sides and corners over a pole in the center. Like a circus tent in a pop-up book, it sprang to life as we all hauled as a team. We aligned our sling-type foldout cots, also in olive drab, in two rows on either side. We slept heads toward the canvas and feet toward the middle aisle. At the base of each cot, a footlocker held our belongings, all arranged by regulation.

Once the inside was in order, we set about policing the area around the tents picking up trash and even single blades of grass. We joined the men to set up a mess tent for meals and cadre tents for the upperclassmen's quarters. The men's tents were within squadron areas marked by giant wooden signs: "Bad News Bearcats" or "Jaguars, King of Beast." Like in the cadet area, all the women's tents were together, separated from the squadrons and unheralded by any signs or mascots.

In Jacks Valley, a *New York Times* reporter, Grace Lichtenstein, and

4. Pugilist

Male Basic cadets outside their tents in Jacks Valley. The men were grouped by squadron. Women's tents were separated. We all slept on cots. Footlockers at the end held our folded issue clothing (courtesy John Visser, USAFA '77).

her photographer followed us everywhere for a piece for their magazine. One woman cadet was half the cover with her chin jammed against her neck, her face hard and serious. The other half was the head of a snarling upperclassman, his helmet touching hers, his eyes locked on hers with a caption of "You make me sick, Basic."[2]

Gail Benjamin was the star as the hometown girl from New York City, and the article included all 15 of the women in Invaders squadron. Gail was a star, though, and deserved to be. She added a New York City attitude to every endeavor, and that gave us spirit. "Misery, misery" was the chant to which many cadets marched throughout the valley during that second BEAST. One cadet in a formation would begin the chant, and the others repeated the words, all using the beat to keep marching in unison. Gail adapted the traditional words to include high-heeled shoes and hair that used to be, no longer in existence. "Misery, Misery, ... Misery, Misery. Used to wear those miniskirts, now I'm wearing fatigue shirts. Used to wear those platform shoes, now I'm sporting combat boots. Used to love to fix my hair, now I can't find it anywhere. Misery, misery, whoa, whoa, whoa whoa...." She was a character fit for the magazine.[3]

So we had this correspondent in our midst, and she was always asking questions and writing things in her notebook. Her cameraman zoomed in while the upperclassmen were in our faces. She was a giant neon sign that followed us everywhere as if we were saying, "We think we are special." Lichtenstein mentioned in the article that upper-class cadets "and even some of the male Basics as well were jealous of all the attention the women are getting from the press, the public and the school." She even quoted an officer in the piece, "If I ever come here again, I want to be a female, because then someone would care about me."[4]

I'm not sure the press cared about us. Their presence didn't make the cadre care about us any more. We didn't ask for her or any of them. In some respect, I enjoyed the attention, but the backlash from any publicity made our lives more difficult. Some of the women, for example Ginny Caine from right there in Colorado Springs, got more attention from the press than many others. "The Caines, residing in Colorado Springs at the time, were overwhelmed by the national media persistent upon following them around telling the story making headlines across America. [In high school] Ginny was followed to her classrooms at school and to the gym while she practiced basketball. She even had news media ride in the family car when she came to the Academy."[5]

Required arm hang for women, pull-ups for men before meals in Jacks Valley during Basic Training. Peggy Walker (on bar), June Van Horn below her (courtesy Cadet Media, U.S. Air Force Academy).

Marianne Owens and Kathleen Bishop were in the *Washington Post* in February 1976 as some of the first women to receive offers from USAFA.[6]

"The *Washington Post* article was a disaster," states Marianne.

4. Pugilist 65

"The female reporter kept asking me about my motivation for attending USAFA, inferring that it was going to be a great environment for women since only 10% of our class would be female. I deflected her innuendos, stating that I was going for the academics, the leadership opportunities, and the chance to serve my country. I did a lot of pushups during BCT and later in 9th squadron, upperclassmen surrounding me and asking if I was happy about the number of males now."[7]

Peggy told me years later of how news crews from the Chicago stations had come to her house to interview her. I think they followed her to the airport on her day of departure for the academy too. So the trail of the press was long and stale for some of the women, but still, every article and every reporter that followed a cadet in a skirt made times more troubled for us there. Solo pictures of the press even made it into our BCT yearbook. Gail was taken to the Gimbel library on the top floor of the cadet library for interviews during the first part of BCT. The academy describes the Gimbel collection as "an amazing array of items pertaining to the history of flight. Five-thousand-year-old seals carved from semi-precious stones and books dating from the 15th century are housed there."[8] Graduates have told me that they did not even know it existed, and as a cadet, I never saw the space. The librarian remembers how the upperclassmen seethed in anger while they waited to escort the female Basics like Gail back to their normal duties following some interviews. "We've never even been allowed to go into that room," they groused. And Gail herself was not enthused because she was never even briefed about interviewing and felt somewhat set up to be facing the press in any room.

I'm sure it didn't help my newsworthiness that I started out with shin splints in Jacks Valley. I was tiny too, usually in the back of an element column or squadron formation. Michelle Pompelli and Tanya Senz were as small as I but certainly more physically fit. An article from my hometown paper announcing my acceptance to the academy reported that I was running every day to be physically ready. There was no ready. Unless I had run 20 miles a day with a clown's outfit on with all the rival high school's football players keeping pace in a convertible Mustang, beautiful blondes sitting next to them as they laughed and taunted me for foaming at the mouth, I would not have really been well prepared. The blondes were not at USAFA, but I still felt as if the upper class thought we were a subspecies, *Homo subsapien*, as we ran full of mud, greasy hair.... *Misery, Misery....*

Through this Jacks Valley phase of BCT, we tackled the assault course with bayonets affixed to the ends of our rifles and had to yell, "Kill, kill, hate, hate, murder, murder, mutilate." We charged around, parrying weapons aside and jabbing dummies positioned along the course. I had never been so tired. The dust excited from its rest by the pounding of all the boots rose in clouds to cover everything. The black boots morphed to a light brown before I pounded even 20 paces from the tent. Dusty sweat covered my face and the sleeve of my shirt when I took the chance to wipe a bit off. Red clouds from smoke bombs set an eerie, haunting stage for our training.

We learned unarmed combat and threw classmates to the dusty ground and jumped toward their jugulars to learn how to disarm and disable the enemy. I liked that exercise. "I like throwing people," I told a reporter, smiling.[9] There was something equalizing and encouraging about learning how to fight back. With the right leverage, I could toss some of those smaller football players over my back. We played soldier and many women showed their strength and abilities. We may not have been physical fighters before. We had not played football or been wrestlers or boxers in high school. But now we were all fighters for our place at USAFA. Men and women trained together on the dusty open areas. "First time I have my arms around a girl in four weeks," one of our classmates told a reporter, "and I'm trying to strangle her."[10]

There were dilemmas with these physical competitions because of the presence of women

"Basic Cadet Utley, Invaders, sir." "Yes, sir." "No, sir." "No excuse, sir." "Sir, may I make a statement?" I am certainly speaking one of those responses here. Second part of Basic Cadet Training (BCT) in Jacks Valley, July 1976 (courtesy Barry Staver Photography LLC).

in the class. The differences in strength and height often made the days more challenging for the women. The average male was 5'10" and the average woman 5'5". The men weighed on average about 155 pounds while our

4. Pugilist

"Hold on to your rifle, Basic." If they wrested it away, you had to carry a stick and endure much harassment. We all went under the barbed wire at the assault course, men and women, side by side during the second part of BCT, Jacks Valley (courtesy John Visser, USAFA '77).

average was 128, and I was no more than 105 after the diet of sitting up like a robot at meals and the real boot camp workout of the daily routine.[11] There were also the key differences in experiences to consider. "For the most part, women had not spent their childhood playing war, climbing trees or swinging on ropes."[12]

In the planning for our arrival, though, Colonel McCarthy, the officer in charge of making the plans fit into the reality of women at USAFA, and his planning team made a noteworthy observation in Texas during some training that included women. "We saw women completing physical exercises and the confidence course for example at Lackland [Air Force Base], but their performance was not very good. Our observation at that time was [that this was] because they were putting women through these programs separately and that they were not getting the reinforcement that men and women going through these programs together might give them. In other words, if the first woman fell off of an obstacle into the water, that became acceptable behavior for all of the women. In our observation for our own program, if you sent men across first, then the competitive attitude and the desire to be just as good would be generated in the women and there would be less failure. And as I've said before that's precisely what happened."[13]

They were right. But there was no need for the qualifier of men first.

The presence of our classmates, men and women, was a very encouraging and motivating factor in my success. I ran after the men watching how they climbed obstacles. I followed some of the women climbing up the cargo nets, hands on the vertical ropes and feet on the horizontals to keep fingers and boots of adjacent climbers from connecting. The men of our class followed us too. At the finish line of the challenges, all of my classmates cheered and encouraged. Any available arms caught the staggering bodies of classmates crossing that last dusty step and offered canteens or shoulders for support. Photos of women holding up male classmates who seemed completely spent made the papers. Many instances of men holding up female classmates occurred and didn't make the papers as often but still created a bond for us. As a team, we defeated the obstacle course, the assault course, the anger of the cadre and the Beast of BCT.

General McCarthy tells the story of the obstacle course to explain the value of training the female ATOs before we arrived.

"Keep in mind," General McCarthy explained, "the acceptance of the class of women was our principal concern. The ATOs went through everything cadets did, and so we had learned from their experience."[14] He continued, "The men were putting the ATOs through the confidence course, and I was standing there at the water hazard with the Superintendent." The obstacle course was a series of log-built walls, cargo net challenges and overs-and-unders that we ran in Jacks Valley, wearing full fatigues and combat boots.

Jacks Valley. Basic cadets on the Confidence Course. Male and female classmates completed it together. Cadet Malkovich, Cadet Diane Moyer (courtesy Clark Special Collections, McDermott Library, U.S. Air Force Academy).

4. Pugilist

Getting support from our male classmates at the end of the obstacle course. We all helped each other. From left: Kevin Keith, Catherine McKenzie, unknown (courtesy Clark Special Collections, McDermott Library, U.S. Air Force Academy).

The water obstacle was a Tarzan rope hanging over a pool of muddy water. General McCarthy continued, "The first ATO came down, grabbed a hold of the rope, made it halfway across and splashed into the water. The next one came out and almost hit her head on the far side (then) into the water." After a whole group of wet ATOs lost to the water, he canceled the exercise and drew the staff together to find out the cause of the failures. Everyone had a different theory for the misses. The athletic department thought it was a factor of upper body strength. Lower center of gravity was one of the rationales, probably from the academic or engineering consultants. Finally, one of the female advisers began, "Well, I have never swung on a rope before, but...." General McCarthy had his answer. They took a group of men out and asked them how many had swung on a rope before. Many had. The ATO women had never. There was a direct correlation between previous rope swinging and success. All the ATOs and subsequently all of us women were introduced to rope swinging before we met the obstacle course or the muddy water.[15]

But it was not all preplanned, measured and then taught. Marianne recalled her meeting with the water obstacle in a different way. "I remember we were at the obstacle course as Basic Cadets, and you were supposed to grab this rope and jump Tarzan-like across this water. My first two times I had fallen in. And I thought, 'I am never going to do this.'" Her cadre member recommended, "Throw your heart over and your body will follow. So, throw your heart over." Marianne eager to learn and to complete the obstacle, thought, *Okay, okay!* "I went running," she explained. "I grabbed it [the rope] and threw my heart over, and I landed on the other side."[16] Sometimes our successes were just pure heart.

I needed heart, as I was not the most physical specimen, so I threw my heart over, through and into the mud sometimes. I panted just to make the run to the challenges, let alone muster the energy and resolve to be enthusiastic about maxing the score or minimizing the time. I was just happy to make it to the finish, to my classmates cheering and encouraging each other. Some upperclassmen became enthused and cheered for their charges, their squadron's Basics, to really throw themselves into it. Their job was to train and through the dirt and wet of the valley, they pushed us physically until heart was all I had left. I don't know if I would have found the heart to keep going if the physical struggles didn't expose it.

I tired from those long, fatiguing, dusty days. The time in the tents luckily was a little more relaxed. June Van Horn describes sleeping in the tent: "My cot was in the corner. Somehow it had slipped out during the night and my legs were outside the tent. It rained all night, and I never woke up."

Motivational movies under the stars still put us all to sleep, but the black sky, filled with shooting stars, took our minds away albeit for a moment. One inky, velvet night, I was on patrol duty with June. We were supposed to yell, "Halt, who goes there?" to anyone we encountered. We twisted our faces into masks of seriousness and then giggled into our palms as we practiced saying it to imaginary enemies. But the night was empty of intruders and full of shooting stars and hooting owls. We sat on the fringes of the encampment enjoying the land that we had not been allowed to even look at during the day. Desperate for sleep, we enjoyed the required alert time more than the rest because of the beauty of the setting. The same fields, covered in a cloud of dirt and a cacophony of commands during the day, were silent and peaceful at night. I did not mind patrol duty. I loved it. It refreshed me for another dusty day.

We had other respites too. The chaplains held a Fourth of July picnic during our time in 1976. We relaxed and enjoyed each other's company for half a day. We rode bikes and talked with other Basics and not just about training. I had the chance to ask some where they were from. I smiled seeing upperclassmen only at the fringes of the fields. We also attended an air show and toured aircraft on display to refill our motivation tanks. I was very motivated by not wearing the long-sleeved fatigue shirt. Drinking ice-cold Cokes with classmates, now friends, shot my enthusiasm skyward.

4. Pugilist

We took some flak for the picnic later as cadre complained how "the standards were lowered" and "your class is weak. No other class had a July 4th picnic." Many blamed that on the acceptance of women. I think they forgot that the bicentennial of our country maybe played a role in the sanctioning of the celebration.

* * *

I remember the day of the pugil stick competition as one of the only during Basic Training that I approached with almost no fear. I was generally nervous every day; I rarely knew exactly what awaited me. Passing upperclassmen commented, "You ladies are going to face the pugil sticks tomorrow," but their tones and faces did not reveal if the activity was something to dread, like the assault course, or a quasi-fun activity like evening movies. Most times, it seemed the upperclassmen spoke in the same rough manner, no matter what they were talking about.

I had no idea what a pugil stick was. Since arriving at USAFA three weeks ago, I had used bayonets to "stab, thrust and jab" the dummies on the assault course and fired an M-16 rifle, completely killing a paper rabbit. But "pugil" was new jargon without any bad connotations, at least to me. I had never heard of a pugilist or seen their battered faces. Probably a good thing. Thanks to my naivete, it was one of the days when I really was unprepared and unafraid.

It started out like any other day in our encampment in Jacks Valley, the notes of reveille screaming from the olive drab, trumpet-shaped loudspeaker mounted atop a telephone pole. I jumped up from my cot. With eyes still squinting, I got on the uniform of that day, of every day: full fatigues and combat boots. This morning, though, we were not "under arms," a rejoice-worthy relief. Those extra ten pounds of rifle accessory felt more like a hundred-pound weight before the blood had even begun to reach our achy muscles. Another reason for my lack of foreboding: how tough could a day be if we weren't lugging our rifles around?

We straightened the sleeping bags on our cots and checked the footlockers at the ends leaving the contents all folded to regulation sizes and in their strictly regulated arrangement. Any morning might also include a lesson in rifle stripping and cleaning or a timed trial of our proficiency in assembling the rifle, its pieces laid out on a canvas bag challenging us. These morning games became easier because of their routines and our understanding of the expectations. This morning, the unknown was the second run over to the area near the assault course for the preliminaries of this pugil stick thing.

The muscular cadre member who got our attention at the end of a run to the assault course with a yell of "Eyes on me, Basics" was dressed in what now was familiar garb of the trainers. We stood at attention, focused on the strapping member of the Class of '78 wearing his fatigue pants and high-gloss black combat boots. Instead of the long-sleeved fatigue shirt that choked our necks, his white issue T-shirt tucked smoothly into the blue web belt on his fatigue trousers. In large Air Force blue letters his last name stood across the taut front above USAFA letters.

"This year, you will be competing for the title of Big and Little Bad Basic," the second-class cadet broadcasted in a booming tone as if his voice box contained its own megaphone. Apparently, the champion last year was crowned the "Meanest Mother of the Valley."[17] His eyes rolled and he smirked toward the other cadre members. Another change in the cadet dictionary to blame on the acceptance of women.

He certainly looked like a big bad something or bad mother or whatever he wanted to call himself.

"This title is a great honor bestowed upon the winner," he told us. "The pugil stick competition will be one of the only opportunities for you to earn any title or respect as an individual during Basic Cadet Training. Up until now, most focus has been on accomplishing and competing as a team. Today you can show us your personal strength and toughness."

A cold, creeping doubt about this day stole in along my seams. *Wait*, I yelled to myself, *I like teamwork*. Teamwork had its advantages. Others helped you out; others cheered you on or dragged you along. It was easy to camouflage your lack of capabilities within the strength of a crowd. I didn't want to play a solo game.

"You will all participate," he said, as if he read my mind.

There were many eager Basics ready to jump in the fray even if it had been voluntary. There were many with grudges that would have assured a great fight if the cadre were the opponents. There was a myriad of possibilities that would have made this day interesting without my having to participate. No escape, though. I knew that. It would be another compulsory activity to endure and survive.

"We will now divide you up into weight classes. Form smaller circles around the cadre members in each area," said the giant cadet as his final directive to the entire group.

Of course, the caveat clarification for the women sneaked in. We would only fight each other. *Fine with me. Solo was bad. Fighting the men would be very bad. The other ladies would be tough enough.*

We arranged ourselves in a circle and the white-shirted upperclassman for our group roughly pushed two women cadets on the shoulders.

"You two, step out here in the middle. How much do you weigh?" he asked one.

4. Pugilist

The pugil stick competition. Cadre (black helmets) with bullet bandoliers refereed and goaded the Basics (men in white helmets) to fight (courtesy Cadet Media, U.S. Air Force Academy).

There was no room for feminine modesty about weight in this venue. "One hundred nineteen, sir," she quickly replied.

"And you?" to the other.

"One hundred twenty-five, sir," she answered, also without hesitation.

They looked about evenly matched and neither dominated in height. I knew one was pretty athletic. She was also more relaxed than most. A few of her brothers were upperclassmen, and she must have been well briefed by them as to what to expect.

Our cadre member handed them white helmets with a grid-type white metal face guard to don. He stepped away from the group for a minute. One jokingly held her hand up like a prizefighter. The other took a quick, quiet bow. Then they both stood nervously waiting for our cadet in charge to come back.

"Quit smiling and put your mouth guards in," he yelled to announce his return to our circle.

"This, Basics, is a pugil stick," he said, holding up what appeared to be a Q-tip for some ogre out of a Grimm's fairy tale.

The color: government-issue green, of course. The ends: padded cylinders with green Naugahyde covering. The shape: Q-tip-ish but big, probably five feet from tip to tip with the center a wooden pole of about two inches in diameter.

He laid another weapon of the same size and color by his feet on the dirt. The pugil stick was now certainly defined and had substance.

"You Basics keep your mouth guards in," he bellowed. "You are going to need them."

"Now, each of you"—motioning to them—"hold one of these while I explain how this match is going to work. You will be paired up with a classmate of about the same size. When you hear my whistle, begin fighting using the padded ends of the pugil stick. You will continue to fight until you hear the whistle sound again. You are not to kick your opponent and you must keep both your hands on the pugil stick at all times. Are there any questions?"

Why in God's name are we doing this? What is the point? I don't want to. Do I have to? Are there pinch hitters? Can I forfeit? Is anyone listening?

Of course, I asked none of these questions, nor did anyone else ask anything. The whole procedure had either been explained so well that it was crystal clear, or more likely, we knew that any questioning would be futile. There was no escape. We should avoid any further delays and just get the whole procedure over with like ripping a bandage off. *Let's just do it.*

We watched as our classmates went at it with the giant fatigue-green Q-tip sticks. Tiny nervous giggles escaped some who watched as pugil sticks missed their marks and swung short or high. The combatants spun around as if in a comedic skit. I hoped that the women were purposely avoiding each other trying to delay long enough to run out of time for their bout.

No chance. Cadet Mr. Big Shot in Charge let them have it. "You fight, Basics, or I will be your next opponent."

Scared into believing that, knowing he would hit them harder than they could or would hit each other, they landed some swats and batted each other around a bit. It was no heavyweight bout. No noses bled. No limbs broke, but both women panted for breath as the time was called with the blast of the whistle.

The fighting continued two by two until my squadron mate and buddy Gail and I were the only two women left.

"How much do you weigh, Benjamin," came the dreaded question.
"One hundred eighteen, sir."
"And you, Utley?"
"One hundred five, sir."
"Okay, you two are next."

Wait a minute, I thought. *That is not a match. That is over a 10 percent difference. Plus, she is taller than I am. This is not fair!*

But fair was not one of the criteria. We squared off with helmets on. We stuffed our mouth guards in, filling our lips with plastic stiffness and our heads with dread. In our hands, we held the pugil stick weapons. I don't know why I worried. This was Gail across the dry dirt-circle from me. We were friends and squadron mates. We were going to have an easy dance across from each other and then be done to sit and watch the rest of the show. Gail was thinking the same thing; I could see it in her face. They are not going to beat us with this game, and they are not going to make us beat each other.

She swatted at me with the giant green Q-tip, and I did the same back in her direction. She bopped me on the shoulder. I did more or less the same to her, hitting her legs or her side.

Then the cadet in charge started his screaming with the same admonitions that the others received. "Get to it, Basics, or you'll be running the assault course all night with your rifles." But I knew, or hoped, it was just to scare us to get going. Gail and I weren't going to play that game. Just as that thought finished forming in my brain, the giant Q-tip, now turned pugil stick, struck me across the side of the helmet. My head bounced left to right trying to steady itself. My eyes rolled trying to find the horizon before my body found horizontal. My teeth flew apart and the blue mouth guard landed somewhere in the dirt.

What happened? This guy just whacked me to prove his point. I thought he was bluffing. Now he has used me as an example and as a punching bag. Finally focusing, I looked to Gail for sympathy and saw her eyes asking instead for forgiveness. *Her* pugil stick was recovering from the blow that she, and not the cadet in charge, dealt me. My eyes, just slowing from the roller coaster of confusion, looked at her askance, *"How could you?"*

But I knew the answer. She had to. We had to. This was part of becoming a cadet, a rite of passage and there was no escaping it. It was part of our training to become cadets, part of the heritage of Jacks Valley and of USAFA. Better to do it, not ask for any leniency, and get it over with.

The strapping cadre member was in his glory now. He was cheering Gail on with "That's the way, Benjamin" and urging me into the fray with, "Now what about you, Utley? Are you going to participate or stand there and get beaten?"

"My mouth guard, sir," I whined, seeing it lying in the dirt and knowing that according to the rules, it must be in place. *Maybe take a second or two to pick it up and some time to wipe it off,* I thought. A little stall before we would have to go back to this inevitable duel.

"I'll get it for you," he said, but instead of reaching down for it, he stepped on it, grinding it further into the dirt.

"I think it is over here," he said, but his smirk betrayed that he knew it was under his boot. After pivoting on it, he picked it up and handed it to me full of dirt. I held the mouthpiece, clenched tighter and tighter in my hand, until the plastic bit into my palm. My teeth bit into my lower lip with the same rage. The anger in me was beyond what I thought I could generate. *What an ass!* He knew exactly what he was doing. He was trying to make me cry or, better yet, trying to be the final straw that would make me quit. *How much more of this can I take?* At that minute, I longed to walk away and never look back. *Screw this whole place and everyone in it. I don't need this crap to have a complete life. I'm done.*

But I couldn't. My anger held my feet in place. *Darn it! I was going to quit when I was good and ready and not a second before.* This particular second, all I wanted to do was stay right here, pick up this big green pugilist weapon and beat this guy about the head and body. *Squash my mouth guard into the dirt and smile about it, and I will get tougher not weaker, you big oaf.*

Seeing me attempting to brush some of the dirt away, he glared and said, "Put it in your mouth, Utley, and get back at it. Good job, Benjamin."

The anger rising up so that my face burned hot, with every muscle in my body tense and hard, I forced the gritty mouthpiece back onto my teeth. Gathering as much dirt as I could in a mouthful of saliva, I spewed out what I hoped was a defiant gob of spit onto the ground. It barely missed my own boots.

"I'm ready," I mumbled. As soon as his whistle blew, I sent the right padded but now anger-powered end of that pounding stick to the side of Gail's head before she even knew what was happening.

There! We both were hit, and we both used the mighty pugil stick to batter the other. *You win, you giants of body and midgets of mind,* I thought. *You got us to do what we wouldn't or couldn't. Now we are one of you.*

We continued the fracas and landed a few more blows each. Gail, stronger and taller, made a more visible impact. But I stood, not my ground by any means, but at least continued standing until the end. I felt beaten. My arms hung loosely at my side, fatigued and limp. My boots felt as if the soles had become rooted in the dirt; it was so difficult to raise them for each step. Even more, I felt pummeled mentally. We had been driven to do what we wouldn't. We joined the fray to join the cadet wing, and it felt pretty lousy right now.

4. Pugilist

After the pugil stick fight. The hug of victory, of defeat, of friendship, of the fight for acceptance. Gail Benjamin (back to camera), Kathleen Utley, July 1976 (photograph by David Cupp, *Denver Post* Collection, Getty Images).

The oversized upperclassman grabbed an opposite arm of each of us and pulled us to his sides. "I declare the winner of this bout," he yelled, holding our arms stiffly against his sides, "Benjamin." He pulled Gail's arm high into the air with his and released mine from his grip as I was released from this torturous event.

That's okay with me. It's over. I finished. I was still breathing. Heading for my place at the edge of the circle, I felt a hand on my shoulder and could only think *"What now, idiot?"* I turned to see not the cadre giant but my friend and the victor Gail. She enclosed me in her arms. She hung her sweaty head over my sagging shoulders.

"We did it," Gail said molding her body to mine with a sigh. Immediately, I held no grudges. We *did* do it. We didn't like it or seek it out. We did it, though, and showed ourselves worthy of the acceptance we sought. We would compete. We would continue. We would console and support each other as we marched through USAFA. That hug holds me to this day. And that friendship, no pugil stick strike or giant crushing boot can defeat.

5

Beanie

"If everyone is thinking alike, then no one is thinking."
—Benjamin Franklin

During our Basic time, we were not allowed any contraband, but letters from home were highly encouraged. I received almost a letter a day and thought that my family was so incredibly wonderful to write to me so often. They were, but only I found out later that my family received correspondence from USAFA asking them to write, letters informing them that it was essential to a cadet's morale that we receive encouragement and support from home. I got it. Plenty and plenty of letters.

Someone sent my dad a picture of the hug after the pugil stick competition cut from a Denver newspaper. "Did you see the picture of you hanging on Benji's shoulder?" he asked in a letter. "You were utterly exhausted, and it tore my heart out to see it, but at the same time I was so proud of you. You well know if you did decide to chuck it and come home, you would get the same loving welcome. It isn't fear that keeps you there; it's your own pride and heart that keep you in there punching."

In our letters back to those at home, we could not solicit any kinds of food. That was strictly taboo. No way. Not allowed. When "mail call" time arrived and we all crowded around the delivery cadet, we knew not to expect any packages. Even if sent, none would make it through the process of calling out the names and passing items back through the group. Cadre held on to anything beyond a normal letter. Irregular parcels meant an interrogation with an upperclassman. Even letters that were inordinately bumpy or thick, the upperclassmen opened and examined on the spot. Any contraband became the property of those with the highest cadet rank, and none of us Basics had any rank—no level above the ground. A little harassment came with the letters that smelled too highly of floral perfumes. Joshing accompanied any letter with SWAK, or the like, written on the outside of the envelope.

The most grievous haranguing came when anyone received a

5. Beanie

full-blown boxed package. For me, one evening, it was a box from my dad. Standing in a dirt circle, gathered around the upper-class cadet who stood on an overturned box, I waited for my letters. "Utley," the upperclassman called out, "you are not listening to the rules. You cannot have this," he said, holding up a small box wrapped in brown paper. He reprimanded me for having food sent, for trying to be sneaky, and for thinking that my family and I could outsmart the cadre. The upperclassman smugly opened the box. His attitude changed when he saw that it was nothing but a red-and-white Mead box of pre-addressed, pre-stamped envelopes to facilitate my writing home. I was glad to have escaped punishment. Still, I felt unreasonably disappointed that there were no treats in the box. Wasn't it better to get nothing rather than see the upperclassmen enjoying slices of home meant for me? I returned to my tent consoling myself with "it's the thought that really mattered."

Later, showing the other women my sensible "gift," I found a stick of gum in each envelope.

* * *

One of the major barriers to the integration was the double standard sometimes perceived and sometimes actual in our treatment. The women lived in a separate area on the sixth floor that was referred to as "the Penthouse" and sometimes "Shady Rest." Our male classmates were rarely allowed up there and most of the upperclassmen avoided it like a plague ward. The ATOs reigned and sometimes interpreted rules in a much different way than the upperclassmen or even the male officers. The double standard first revealed its ugly self to me when my sister came to the academy.

After we returned from three weeks in Jacks Valley and prepared for the formal acceptance into the cadet wing and for the beginning of the academic year, my sister stopped by the academy on her way home from the same vacation that would have brought me to USAFA six weeks before. I would have taken a ride back at that point even though I had refused the ride out. Being late was no longer the issue.

Marching to the noon meal formation, at the command of "Eyes right," I snapped my head to the forty-five-degree angle to my right. I had done it so often by now; I did not even have my brain engaged. Marching wasn't so bad. No one talked to you directly, and unless you fell flat on your face, you could go somewhat unnoticed for the brief time it took to get around the quarter mile of terrazzo. With my body marching forward and my chin pointing somewhere between my sternum and shoulder, I stomped along, daydreaming about the end of Basic Cadet Training. It was almost over. We were back in the cadet area for the final wrap-up and acceptance. Acceptance. What an odd term for graduating to the second lowest level at the academy. There wouldn't be any new Basics until the summer of next year when the Class of '81 arrived. Until then, we would still be at the bottom of the totem pole, but we would no longer be "Basic cadets." We would soon be "fourth-class cadets." The ramifications were insignificant, but at the time they felt like huge, glorious freedoms. They were definitely hard-earned.

As I marched along with my eyes facing right, a flapping red, white, and blue coat caught my eye. Sitting on the coat, atop the wall of the Cadet Chapel was a little blond girl and nearby someone yelling, "Beanie, Beanie." I was so stunned that I couldn't think. My family. Only my family called me Beanie. I could barely keep my feet moving to not be trampled by the cadet behind me. There was my little niece, Cathy, who lived in Florida. I knew that they had been camping out west since I had started, but when did they get here?

I totally missed the "Ready front" command as my eyes stayed locked on little Cathy, still stunned. A poke from my classmate to my left broke the spell. I brought my eyes back to front about 50 paces after the rest of the squadron. I was confused. My heart pounded harder than it had during the most grueling run. *My sister was here, close.* I hoped that somehow, I would be able to see her. A sweet voice from home, a blood relative who would not correct my posture, ask me about any current events or give a hoot about who was in my chain of command. Maybe we could even sit down and eat something, anything, a candy bar from the vending machine, without bracing in the rigid posture of the Buckingham Palace guards.

Don't get your hopes up too high. Ground level after this hole we were living in would be plenty of a break. Just a little time to say hello and some hugs would be enough.

It seemed like it took forever to reach the doors of Mitchell Hall, where we were dismissed from the squadron marching formation. Upperclassmen, as always, free to walk where they wished and chatting with their buddies, headed for the tables by the most direct route. I herded to the right-hand wall of the dining hall with my fellow Basics. But as they angled off toward the tables, I was dawdling and gazing around, an absolute taboo. I needed help or advice. *Where would the most likely source be?* In our air officer commanding (AOC), the officer in charge of Invaders, I found an authority who I thought would be able to help me.

He was talking to another cadet, an upperclassman, so I would have to wait my turn. If I stood any chance of gaining permission, I needed to follow all protocol. Dutifully, I trailed behind trying not to be conspicuous enough to attract attention from any others as an errant Basic needing extra training or questioning. I needed my AOC, the major, to notice me and give me an opening to ask my questions. Unluckily on that day, our squadron tables were as far into Mitchell Hall as they could be, away from the chapel wall. By the time we stopped and I got acknowledged, I knew I already lost precious time to find her, to talk to her, to escape.

"Major, sir, may I ask a question?"

"Yes, Utley. What is it?"

"Sir, my sister, my big sister, my sister from Miami, well she's…."

"Utley, calm down take a deep breath and just tell me what is happening."

By this time my shaky voice and watery eyes threatened to cost me even more time. I concentrated, trying to expand my lower lid and tilt my head back to keep any tears from falling but still keeping my eyes on the major. *This is a good thing, God. This is a reasonable, rational request. Please, please don't let me cry.*

"Sir, as we marched by the chapel wall at the eyes right [trying to explain that I would never have been looking around if it weren't in direct response to a proper command] I saw my sister and my little niece on the wall. I knew they were camping out here somewhere, and if it is at all possible could I go and talk to her, just for a minute? Please, sir?"

"Go, Utley. That's fine. This is virtually the end of Basic Training. I don't see why you shouldn't have some time with you sister. Make sure you are back by 1330 and let your table commandant know where you are going and that you have my permission."

"Thank you, sir. Thank you. Sir, thank you very much."

"Go on, Utley. Get going. Get out of here."

"Yes, sir. Thank you, sir."

I moved as fast as anyone could, running while pretending to walk, keeping against the walls and squaring all corners until I got to my table.

"You are late, Utley," declared a sophomore, cadet third class or Three Degree, that I thought of as "Flame." It was an appropriate nickname, as nothing but flaming, harsh remarks would ever come out of his mouth. "Give me next Air Force Day."

"Sir, I was talking to the AOC."

"Utley, I did not ask you where you were. I don't give a damn about where you were. You are late to my table, and I want to hear the next Air Force Day." He screamed moving his face to within inches of mine.

"Yes, sir. Sir, the next Air Force day is…." *Man, I can't even think what day it is today. Okay, this is the beginning of August, so the next Air Force Day is…. I don't know…. Should I make one up? Will he even know? I can't afford to take that chance. My sister is out there and may not wait long. I racked my brain and by some mini miracle it came to me, even through* the blinding pressure. The training paid off. The rote words from my *Contrails* came into focus.

"Sir, the next Air Force Day is August 6 commemorating that day in 1945 when the atomic bomb was dropped on Hiroshima, Japan. The bomb, nicknamed 'Little Boy,' was delivered by the B-29 *Enola Gay*, under the command of Lt. Colonel Paul W. Tibbets Jr."

The calmer voice of the table commander, a very mellow member of the Class of '77, sneaked in right after my last syllable to prevent the "flame" from reigniting. "Okay, Utley. Don't be late again."

"Sir, may I make a statement?"

"Sure, Utley. What?"

"Sir, the major gave me permission to go over to the chapel wall. Sir, I saw my sister from Florida there, and he gave me permission to go and find her. He just said I had to check in with you first."

"You saw *what*, Utley? I think you are having a Basic Cadet hallucination. Are you sure, Utley?"

"Yes, sir. I am, sir. I think if I hurry, I can find her."

"Okay. If the major said it was okay, then go, Utley. Good luck."

I took off, hearing some muttering from C3C "Flame" and some other upperclassmen about the women getting off so easy. "Do you remember our Basic Training days when we were never even allowed to eat...." But I was on my way and soon out the door heading across the terrazzo.

The gods were smiling on me, albeit for only those 200 yards. No one stopped me. No one even looked at me while I ran across the marble strips. I would never have "illegally" cut across the center of the area, even now to save a few minutes. *Now I know why they have you running all the time,* I thought. *So you are in shape for those marathon runs trying to get away from here.*

Be there, MaryAnn. Please still be there.

But she wasn't. One older couple taking pictures of the front of the chapel and one upperclassman wearing the escorting dress uniform with white gloves guiding a few teenagers on a tour were the only stragglers left after the spectacle of the parade of the cadets. I'd have to try the cadet for some info. *Semi-safe,* I figured. He would not yell too much for fear of scaring these teenage prospects away. "Sir, may I ask a question?"

Annoyed but on his best behavior in front of civilians, "Yeah, what is it?"

"Sir, I saw my sister here on the march over to Mitch's and I was wondering if you saw her. She is short and she had a little blond girl with her, and her husband is tall...."

"No, I didn't see her," he said curtly and turned away. But with the teens watching like hawks, he must have reconsidered. "There were tons of

5. Beanie

people here during the formation. Head back to your squadron and ask if she called. That could have been the case. Good luck."

"Yes, sir," my automatic response.

At a sprint down the ramp and across the marble strips, I passed directly under that same upperclassman and the teens I had left. His voice from the wall reached my ears around the brim of my wheel cap: "Yes, now she has to stay on those strips of marble. It is a requirement all through these six weeks of Basic Training and through their first year.... No, they don't mind, and it prepares them...."

Get me out of here....

Back in my squadron, the cadet at the desk handed me a note as I ran in from the stairwell panting. "Utley, your sister called looking for you. Said she would check back later. She said they might visit the field house. If it's okay with the major, you can go."

"Thank you, sir. I got permission, sir. Good afternoon, sir," and I was back out the doors and back down the six flights, winded but not as tired as I had been throughout the whole training. I wasn't running through the assault course, to the pugil stick area or to unarmed combat. I was running and smiling knowing in about six minutes, four if my legs would take me faster, I would have someone, three someones, in fact, whom I didn't have to salute, who would accept answers beyond the five responses, and who could just hug me without asking, "May I touch you?"

I made it in record time. I was going to see my sister. Grinning from ear to ear, I breezed all the way down. Most cadets were still at the noon meal, so no one stopped me or even really saw me going.

But my family was not there. *How could they not be there? Where else could they possibly be?* I needed them to be here to see me to recognize me as a sister and an aunt and a human being. *They had to be here. Maybe they were all in the bathroom together.* I looked everywhere.

Finally, giving up, I headed up to the terrazzo with my shoes dragging as if filled with lead and my head bent forward with disappointment. Everyone was streaming to his or her assignments after the meal. I joined the tide.

Later, heading into the sixth-floor ladies' area to get into a different uniform for an afternoon activity, I noticed the ATO sitting at the desk staring at me as if she knew the story. Possibly, it was just idle curiosity. Yet there was something about the way she looked at me. I almost asked for her help, but I was tired, and it was too late. I kept walking with an "Afternoon, ma'am" trying to keep in the sobs until the door closed behind me.

Later, I got the letter with the whole story. The chain of command in my Invaders squadron, from the officer in charge, to the table commander, to the cadet at the desk, had given me permission to see my sister. The ATO surrogate, probably using the regulations and not any chain of command, told my sister no. Living apart from the men, under a second set of rules, had pulled that chance for a little bit of hope and home away from me.

Dear Beanie,

We tried but we couldn't do it....
I called Thurs from Mesa Verde and was told it might be possible to see you. They would know more Sunday.
I figured I might catch you at mass, so we went at 9:00. No girls. No one running. During Mass I found out why. There was a 7:10 mass which you no doubt went to. Just in case, we wandered around till 11:00. We came back up at 12:00 to see the noon formation, and we saw you. I think you saw us too. You were the only one with her mouth wide open. Cathy was sitting on my red, white and blue jacket that was hanging over the wall right where you looked when they said, "eyes right." After you went in to lunch, I reached 9th Sq. It was still encouraging. So we went to the field house and came back. I checked at the desk in Arnold Hall, and they said to call 9th Sq. I did. The guy suggested I try to get you in the ladies' dorm. I did, but the woman who answered the phone said no visits and no calls unless an emergency. I couldn't think of one quickly enough.
Maybe they told you I tried to see you. Maybe not.
Anyway, we were glad to see that you're still here and sorry we didn't get to visit. I only wish they had made up their minds earlier.
I hope you survived Jacks Valley in pretty good shape. Take it easy. We've been thinking of you. N.I.C. You can do it.

Lots of love and kisses,
Pook, Bill and Cathy

P.S. Cathy got a USAFA t-shirt today.

I could have used their encouragement and warmth that day. Then again, maybe I would have left right then and there with them. Maybe my resentment about the double standard of the female ATO who made a negative decision for me, a woman cadet, not knowing that a major and my cadet commander, truly my commanders in traditional academy hierarchy, had already made one in my favor, gave me more determination to stick it out.

6

Guinea Pigs

"It is not easy to be a pioneer—but oh, it is fascinating!
I would not trade one moment, even the worst moment, for
all the riches in the world."
—Elizabeth Blackwell

There had been ATOs at USAFA before these women who came to train us. When the academy was founded, the official site in Colorado Springs was not completed, so the first class, the Class of '59, began their academy days up in old World War II barracks at Lowry Air Force Base in Denver. Some thought was given to bringing upper-class cadets from West Point or Annapolis to train the new Air Force cadets, but it was decided to use ATOs for training and to serve as role models until the first classes moved through the training and years to become upperclassmen for the subsequent classes.[1]

The original ATOs were a select group of men chosen to train the first three classes of cadets at the fledgling Air Force Academy. Colonel Robert "Bob" Hess, one of those first trainers, shared his experience. "I was an ATO, First and Second Lieutenants who acted as upperclass at the interim site on Lowry AFB." These original ATOs trained the first three classes, the Class of '59, the Class of '60 and the Class of '61. After the Class of '61 had completed their summer training, the ATOs returned to assignments throughout the Air Force.

Those men ATOs were all pilots or navigators because all the original cadets attended navigator training while at Lowry. Hess explains that all the ATOs had agreed to "remain single while we were ATOs and they lived in two-man rooms in the wooden buildings, structures from WWII."

The ATOs trained the cadets. They marched them everywhere, to academics, to the mess hall and then back to their rooms single file when they were fished eating. "We basically marched them anywhere or ran them everywhere and we always said, 'follow me,'" Hess shared. "If we wanted them to take a lap around the field with a rifle and bayonet held in their arms, we would do the same. We never asked them to do anything that we didn't do in front of them."[2]

As part of the initial planning for the admission of women starting back in 1972, there was a decision to again use ATOs, this time women to serve as surrogate upperclassmen and role models for the female cadets. In a manner, they were going to be the test guinea pigs before we were the real guinea pigs.

The process for choosing the ATOs had been very selective. From a group of more than 600 records of first and second lieutenants, "eventually 70 ATO applicants were interviewed in person by the Commandant, the Vice Commandant, one Group Air Officer Commanding, and officers from the Special Planning Staff." Like us, they were submitted to many tests with their associated acronyms, including a physical aptitude examination (PA) and the Minnesota Multiphasic Personality Inventory (MMPI). "Final selections were made on the basis of outstanding military qualities, athletic ability, and human relations and communicative skills. The 15 ATOs selected represented five major Air force commands and 11 career fields."[3]

The 15 female lieutenants were commissioned officers in the United States Air Force, what we would be when we graduated. They had completed their officer training, either through a short course called Officer Training School after their graduation from college or had gone through college in ROTC and became officers when commissioned after graduation. They were out in the Air Force, starting their careers as munitions loading officers, information officers or weapons loading chiefs.[4] Through interviews and records reviews, the female lieutenants were selected to come to the academy and serve as surrogate upperclassmen for us, the first women cadets. From January until June 1976, they had undergone an abbreviated course to represent the four-year academy experience. Their five and a half months of training also gathered data for our success. They did their BCT in the winter. They were the women that Peggy had seen in Mitchell Hall during her pre-acceptance visit. It was about them that the cadet had said, "They don't belong here." They were to be our mentors.

On 29 December 1975, the academy announced the names of the 15 women ATOs, ranging in age from 22 to 26:[5]

2LT Dawn M. Reed
2LT Paula A. Gathwright
2LT Elizabeth Goolsby
2LT Irene L. Graf
1LT Charlotte Greene
2LT Susan N. Hamilton

2LT Yardley N. Nelson
2LT Shirley L. Popper
2LT Virginia Procino
2LT Rebecca Ritchey
2LT Ronda M. Roszel
2LT Kathryn L. Sheridan
2LT Bonnie L. Stephan
2LT Rhoda A. Sweitzer
2LT Terry J. Walter

General McCarthy explains the thought process behind the ATO program. He cited the "Role Model Concept" as the primary reason for having female officers as ATOs. "One of our major concerns was the importance of the informal training that goes on here: the upper classes molding cadets in the fourth class to what they think a cadet should be—a very informal process, we don't allocate any training time to it necessarily—it is one of the strengths of the institution."

He continued, "Now when you bring women cadets in for the first time in this previously all male environment, they are going to ask the same questions and they are going to try and pattern their behavior after the men, and one of our concerns that was supported in much of the literature was that women might tend to adopt, either intentionally or in most cases unintentionally, mannish characteristics or mannish responses not that they wanted to be mannish. One of the things that we wanted to stress to the women cadets coming in is that they could accomplish all of these tasks that we expected of them without the loss of femininity, so the role model was introduced to help reinforce this idea of yes, there is a male cadet that she should pattern herself after, but here is a woman living in the same environment who has completed a training program similar but abbreviated to what you are going through. She can do it, and she is a woman and acts like a woman."[6]

McCarthy explained the second important aspect of the ATO program as an opportunity "to teach men how to relate to women in the training environment." He continued, "Probably the most important aspect of all was the fact that six months in advance of women cadets arriving here a small group of women arrived and we put them through a training program that the cadet wing was able to look at and relate to, and they began to make judgments about how women would perform as cadets, which probably created a positive attitude on the part of the cadet wing more than anything else we did, and there lies the true value of the ATOs as far as I am concerned. They were the cutting edge; they were the shock troops. And I remember the first time these 15 ATOs marched across this terrazzo to the noon meal formation there were hundreds of men standing under Fairchild Hall and around, just looking to see how they marched and how we were treating them."

Cadets directly participated in their training. Somewhere between 200 and 300 cadets actively joined in the training of the ATOs, which, according to McCarthy, "changed a lot of attitudes." He continued, "We intentionally selected cadets who had negative attitudes, as well as those that had positive attitudes, and I know of no case where the negative attitudes were involved in the training program that they left with those negative attitudes."[7]

In addition to changing the attitudes of the men, McCarthy explained that another value add of the program was that the administration and the planning committee gained confidence in the plans they had put together.[8]

General Allen, the academy superintendent from August 1974 to July 1977, the two years prior to and the first year of our cadet lives, summed up the ATO program: "I don't think anyone fully anticipated how much benefit would be reaped from the 15 young women we brought in as Air Training Officers. The ATOs eliminated a number of intangibles before the freshman class arrived. They conditioned us to the notion of having women on the campus before the freshmen women arrived and they taught us how women would react to our training programs. I think they were a tremendous asset. It's kind of interesting: I don't think the freshmen girls themselves or the ATOs appreciate how much that approach influenced the way things went. I think it did a tremendous amount for us."[9]

But hearing from some of the ATOs themselves paints a different picture.

Lieutenant Graf didn't really know what was expected of her. She even asked one officer, "Does this mean I get a second commission?" "I knew it was going to be rigorous," she explained. She didn't know exactly what role the ATOs would play, but she thought they would be "disciplinarian for the new girls coming in." She even thought their role might be a separate women's academy.[10]

Lieutenant Graf said she thought the cadets would be accepting of the ATOs. She thought "we would fit right into the wing, and everything was going to be just great." She admitted she was naive about it.[11]

Lieutenant Walter had a different perception before her arrival. "I was under no false impression. I knew that it was going to be tough, and I knew that BCT was probably the hardest thing that a cadet goes through, mainly because of the initial shock of the whole thing." Like Lieutenant Graf, she didn't know exactly what her duties would be, but she said, "I never, never imagined that it would be as a surrogate upperclass. I just could not imagine officers taking over cadet functions. I couldn't imagine cadets tolerating it."[12]

Lieutenant Walter went on to explain that the first week was "just peachy, I mean we were the ATOs, and there were the press conferences, and General Allen was talking to you—I mean, we felt pretty darn important." They got up the first few mornings and had a nice breakfast at the Officers' Club. She got nervous, though, when the ATOs were moved into the cadet dorms knowing the Officers' Club breakfasts were over and the training was about to begin. "I knew that they weren't going to be nice to us for much longer, and that these warm, friendly fuzzy people were going to disappear. I knew what they were going to do to us, but I still did not believe they'd actually do it." The next morning, she woke to whistle blowing and cadre yelling, "GET UP! GET UP! SPEED OUT, ATO!" She explained that some of the noises they heard from behind the closed doors made her laugh. "I walked out with a big grin on my face and I was laughing, and this big cadet who was nine feet tall appeared right into my face with his big eyes and he said, 'What are you laughing about, ATO? Wipe it off,' and I was back against the wall, and I didn't laugh much for about five months."[13]

Cadets did come up to one of the ATOs after their training was over and said it was one of the hardest things they had to do during the first days. "We had never yelled at a girl like that before," they clarified. "I mean, we have girlfriends, and our sisters and our mothers, and to go up to a girl and start telling her how ugly she looks." It took the cadets some time to get into the role of yelling at women. Certainly, the positive motivation philosophy promised was not what was practiced.[14]

The ATOs got yelled at, got oinked at and mooed at marching to meals. A dead rat was left on one's desk. They were treated like female cadets but certainly not as officers. In fact, they didn't fit in with cadets and often not with their officer peers either.

A standard announcement in the cadet area would begin with "Attention in the area, attention in the area" and continue with the rest of the announcement. One graduate remembers the first day the ATOs marched across the terrazzo, the speaker from command post started an announcement: "Tenson in the area, tension in the area." One printed announcement came down looking for an escort for a visiting dignitary "an officer or an ATO." Not a joke since ATOs were officers but not really thought of as such. There were ATO jokes, though. One ATO told one for posterity: "What is the difference between an ATO and a trash can? Trash can gets taken out twice a week." "That," she said, "was the original ATO joke, and they went from bad to worse."[15]

A cadet who was a Doolie while the ATOs trained told me, "We got a directive," not official in the regulation kind of way but the kind that comes from upper-class cadets that have established their influence over others, "to call them 'sir.' One day, I was running the marble strips and passed an ATO, slowed to salute and yelled, 'Good morning, sir.' She dressed me down and told me to call the ATOs 'ma'am.' For a few weeks, I still called them 'sir,' many called them 'sir' when we saw them. That is what we were told to do."[16]

I regret that I learned all these stories only long after graduation. Some of these shared experiences would have bonded me and others with the ATOs. Kudos to them for not telling them, though.

Their jobs mutated when we arrived. "The whole concept changed the minute the girls arrived," an ATO said. It became apparent that women cadets would be able to be trained within the existing cadet-run system. "Our role changed drastically," Lieutenant Walter described. Their job became more accepted as that of a female officer role model and sometimes counselors than the function as surrogate upperclassmen that the original ATOs were.[17] Eventually, the male cadets did all our training.

The women ATOs had broken through the initial barrier at the academy but had received little of the attention and almost none of the credit. Like taste testers for royalty, they got a sampling of what it might be like to be female cadets at USAFA, but once the fare seemed safe, they were pushed aside for us to take their places. They had left their Air Force jobs to be hounded by male cadets of lesser rank and age. They vied for the opportunity. They put their careers on hold to hold a place for someone else.

Marianne later said, "It wasn't until years later that I truly appreciated them for what they did, for having to live through all this, with aggression coming at them from all sides."[18] My classmate Janet Wolfenbarger remembers the ATOs "regularly referred to us as 'Ladies,' in a loud and demanding tone of voice that was not meant to be complimentary. I find it so appropriate that we coined the term '80's Ladies' in reference to our unique group, because every time I use the term, I hear our ATOs' voices." She has come to understand that the ATOs played a treasured role in our journey. Betsy Joviak Pimentel recalls the "one assigned to my squadron, Lt Shirley Popper, was a realistic mix of supportive and focused on training in the early weeks. It took me a long time before I really appreciated what they did." It took a long time before I appreciated them too.

7

Doolie

"It's hard to understand inclusion unless you've been excluded."
—Billy Jean King

The acceptance ceremony on an early August evening changed our names from "basic cadet" to "cadet fourth class" or "four degree." Upperclassmen called us "Doolies" now and "Four Degrees" also. "SMACKs" sometimes, for Soldier Minus Ability, Courage and Knowledge. The list of things that I was called, and the names used for me, grew. I came off the bus as "Kathleen Marian Utley" or "Kathy Utley." I was officially "Candidate Utley" and "Nominee Utley" before that, but no one had called me those names, only written them. "Ugly," "slow," "stupid" and "weak" had been yelled in my face many times over the past six weeks.

The commandant's committee for the preparation of our arrival had not forgotten to consider names for us either. Their plan for the ATOs included discussions about what to call them, settling on ma'am or lieutenant. For us, the plan called for "miss" or "cadet." That was perfect, equivalent to the men's titles of "mister" or "cadet." I snapped rigid at attention and answered to "Cadet Utley," or only "Utley," or "miss" when addressed by the upperclassmen.

I was still "Kathy" to my classmates and roommates. Dad wrote to me as "My Darling Daughter," Mom as "Dear Beanie," and siblings Pat, Mike, Barb and MaryAnn, when they wrote, started their letters with "Bean." I became "Basic" before I even knew what that was and now "Doolie" before I had memorized what that word meant. "Cadet fourth class," or "Doolie," was still the lowest rung.

This was only step one, however, in a long march of four years to the finish. One step at a time. Acceptance gave me four-degree shoulder boards made of solid black felt for my previously unadorned shirts and more names to add to the list on the way to "lieutenant."

It was at least a start.

* * *

"Good morning, sir. Niners," chirped Peggy. We added Niners to our greeting to announce our squadron pride. Her enthusiasm and the bounce in her step were not dampened at all after the rigors of BCT. With her eyes and voice smiling, she added, "Beat Wyoming," a cheer for the upcoming football game against the University of Wyoming.

"Good morning, sir. Niners. Cream the Cowboys," June blared out. June grew up in a patriotic family. Her dad enlisted in the Navy from high school and her brother was a 1974 USAFA graduate. She went to Colorado State University, majoring in pre-veterinary studies. "I grew up on a farm in Iowa and loved working with our cattle. I volunteered with our veterinarian. My hands were small, so when a sow got exhausted in the sweltering summer heat and needed help pushing her piglets out, I was the one who could reach in and bring them into the world." She was always interested in flying. She applied to the academy when she heard that the academies were opening to women, and she also applied to veterinary school to fulfill her dream of working with animals. She was accepted to both. Her decision to join the first women at USAFA was measured and logical. She had reached the maximum age limit to become an incoming cadet. She had one shot at the Air Force Academy. She knew she could apply again to vet school if she left the academy, so she marched up the ramp with our class with the safety net of working with animals still intact if needed.[1] She was strong, undaunted and mature. She welcomed every challenge of BCT from the tower of logs on the confidence course to mortally wounding assault course dummies with bayonets. She was now up for the challenges of the academic year.

"Good morning, sir. Go, Falcons. Beat the Cowboys," Debbie Wilcock chimed in. Like Peggy, Debbie found a way to stay peppy. She smiled most of the time. BCT was over, reason enough to smile. Debbie's brother also went to the academy and graduated in 1975, so she remembers having a lot of info about it. "My parents (both from the military) could not afford to send me to college and told me my only option was to go to the Academy if it allowed women. My brother tried hard to talk me out of it and really didn't think it was a good idea for women to attend. I felt like I had something to prove to my brother, plus I was excited about being a pilot, so I applied."[2]

"Good morning, sir," was Marianne's perfunctory greeting to the upperclassmen we passed on our way to Fairchild Hall, the academic building. Marianne had her game face on. Still determined, still serious, she was about getting the job done and done well.

"Good morning, sir. Beat Wyoming," was Gail's add to the column of

7. Doolie

greetings that we Doolies obligingly spoke to the upperclassmen we passed on the terrazzo. She echoed the words of Peggy, but the tone was different. Gail couldn't match Peggy's enthusiasm on this particular morning. She was focused. She was heading to an engineering class, and she was already mentally reviewing the material. She said the required words and added the expected football team support all matter-of-factly. She met the days, the steps and the words as they came and as her mood met them.

"Good morning, sir. Go, Falcons," I said from the rear of our group of 9th Squadron Doolies. A gust of wind grabbed my flight cap, threatening to fly it over the wall and down to the base of the "Bring Me Men…" ramp. Books in the crook of my left arm kept my right arm free for saluting any passing officer and, luckily, available to hold on to the hat. Greeting the upperclassmen along the way to class was now part of our routine.

Gail, Peggy, Marianne, June, Debbie and I were Doolies in 9th Squadron now. All bound together as a group by six weeks of Basic Training, all from different places, all fourth classmen, all Niner Doolies.

Doolies. According to lore, it's from the Greek "doula" for slave. The same rank as the plebes at the U.S. Naval Academy. Plebes, the common people of ancient Rome. No expectations of any privileges with those titles. We were above Basics, though. No longer required to run everywhere, we were still restricted to movement only along the marble strips at the perimeter of areas. We still sat at attention to eat. We continued memorizing *Contrails* and met with upperclassmen in the halls to verify our learning. Still, most of us had made it. So far, only five women, about 3 percent, and 89 of the men of our class, a little over 6 percent, had given up on the academy and "punched out" as if ejecting from an aircraft. General Allen's prediction for a lower attrition rate this year was coming true. The cumulative attrition would end up being 23.3 percent by the end of our first year, from June 1976 until June 1977. That was 3.7 percentage points below the five-year average. The numbers were better than the attrition numbers for Annapolis or West Point as well.[3] Something must have been working.

We were assigned to 9th Squadron now. From the BCT summer squadrons, all the Basics were split into our Doolie squadrons to join the remainder of the cadet wing for the academic year. The 4,400-strong cadet wing was divided up into 40 squadrons, or four groups of ten squadrons each. Women were divvied up into just the first and second groups, squadrons 1–10 and 11–20, leaving 21–40 male only. We women BCT survivors from Invaders were assigned to squadrons 9 and 19, yet we would still have to live in the sixth-floor women's area. Our male classmates moved into rooms within their squadron areas, living within sight of the upperclassmen.

We completed Basic stripped of all we were when we arrived. No longer valedictorians, team captains, academic or athletic all-stars, Eagle Scouts. We had been Basics and we were now Doolies and began the rebuilding process as part of the cadet wing of the United States Air Force Academy. Now we were going to fill the voids that were left when high school, home, family and friends were removed. My parents and siblings who previously spoke to me and touched me with hugs and smiles now only reached into my life in letters. Other voices began to fill my ears. We gained military knowledge through the words of *Contrails,* the military heritage of our new home. We had figured out some of the workings of the cadet wing and six weeks' worth of an understanding of the women's place here at USAFA.

We had run through BCT to gain our wind for the year ahead. We were more physically fit than when we arrived at the base of the ramp six weeks ago. Our lungs and blood stream had acclimated to the altitude. We walked up the six flights of stairs to our rooms now without panting for breath. We had learned the beginning of the vocabulary of USAFA. When upperclassmen commanded "Forward march," my left foot immediately responded, so I did not start out bouncing out of step. When I heard "Right face," I ended up facing right without having to check direction. We marched as a unit of Basics and now would join all the classes to march as squadrons.

The rest of the cadet wing, who had been away for their summer programs, returned to the academy. Now there would be three times more of "them" around. These returning upperclassmen were seeing the women for the first time. The first month of the semester became an abbreviated repeat of BCT. Extra chaos ensued with lots of additional yelling time for those who had missed out on their chance during the summer. Luckily, it was not as much of a shock to my ears and soul to be blasted just for living or for violating the sanctity of the men's domain anymore.

Sometimes when women passed upperclassmen on the terrazzo, we heard "bitch" from their mumbled voices. Sometimes they didn't mumble it at all. When a line of Doolies trotted on the marble strips and greeted upperclassmen as they passed some of the angrier upperclassmen would call us out by saying, "You bitch, get out here for some extra training." A male classmate told me that once he was running with a group of Doolies on the marble strips when an upperclassman shouted, "Hey, you, Doolie with the sloppy uniform. Get out here." This classmate self-identified and stopped for some extra yelling about his sloppy shirt. He mostly punished himself, however, for even stopping. How did he know the upperclassman really had identified him as the messy one? Was that sophomore or junior cadet making a random accusation just to see who would volunteer? He laughed at being so naive and foolish.

7. Doolie

For the women, it was obvious who was being chosen. Usually, we ran with our flights or with our squadrons or with some classmates, but the ratio of Doolie men to women was ten to one, so we most often were surrounded by men and not in groups of women. June later remarked, "All freshmen were called many 'interesting' names by the upperclassmen. Sometimes, the gals were called different names."[4] When an upperclassman called for the "bitch" in the group to "drive out" of line to be corrected (stop and face the caller), there was no disguising ourselves. Gail tells of how she was sitting at attention at the foot of the lunch table in Mitchell Hall. An upperclassman came up behind her and whispered that no one wanted women here or n-----s. She couldn't report him. She couldn't even identify him. She was not allowed to turn her head to look while at the tables.

I heard that at the Naval Academy, the women were called WUBA. The acronym normally stood for Working Uniform Blue Alpha, a description for a uniform like our "combination Alpha" uniform of blue skirts or slacks and blue jackets. But the male midshipmen used it to mean, "women with unusually big asses" or worse, "women used by all."[5]

"Hey, you, bitch," an upperclassman called one day when I was running the marble strips on my way back from the noon meal. I drove out from the marble edge executing a strong left face in the direction of the upperclassman. I stood stiff at attention and focused a serious but not stern stare at the chin of the attacking cadet. I tried not to look directly into his eyes. Staring straight into the hatred made it hard for me to conceal my charade of indifference or to ignore the burning anger that some of these cadets felt toward any of the women cadets. "We don't want you here," he whispered as if that were news to me. "Yes, sir," I answered. I tried to let it all go. I attempted to look contrite for sins that I never committed, until he or any other of the upperclassmen who put me through similar experiences spent their venom. Finally dismissed, I marched away hoping that the anger they spewed would not regenerate. A tough day today might soften the future days, months, and years for us, the feminine intruders to the academy.

Filling of eager minds began as classes started. Our weekday daytime hours filled with the required courses to make "officers and gentlemen" and university graduates out of us. Engineering courses were a must, no matter your interests. The curriculum required military studies all four years. Physical education was an ongoing obligation. My older sister had told me never to take more than 15 credit hours per semester. At USAFA, I was astonished to find that 21 was a standard load. I tested out of Spanish, thanks to those years in Costa Rica, but that put me ahead into classes with third-class cadets from the male-only Class of '79. Many didn't want women to violate the gender purity of the classroom as we already had in the dormitory halls and on the athletic fields. Those classes were tense and awkward. How did Gail and June who transferred in with so many credits handle the hostilities in so many of their classes that expected to be male only?

Marianne explains how in her math class upperclassmen would walk by and scrape their shoes over the top of hers, then report her to the instructor for having unshined shoes. She did lots of push-ups in math.[6] She admits she chose her academic major because that one instructor did not make her do push-ups in class.[7]

Marianne also tells the story about a similar experience not in the classroom but in Mitchell Hall. The Dallas Cowboys cheerleaders came to the academy. While Marianne stood at attention at the table, the upperclassmen ordered her to clap for the cheerleaders when all the male cadets were cheering and whistling and clapping. "No, sir," she answered. She did lots of push-ups that day too.[8] If their cause was to get Marianne fit, they were accomplishing something. If their intention was to force her out, they were failing.

Gail remembers the classroom as a very isolating place. In one of her classes of all sophomores and juniors, no one would sit with her, no one would do any projects with her. "So, I sat very much alone," she recalls. "I like to sit in the front because I learn better that way. But no one would sit near me…. I felt the penetrating eyes of people looking at me." The result was a determination to do well and her baffling the class with an exam finished in record time, not because she gave up but because she used the challenging eyes to focus her steely determination and ace the test.[9]

Unlike the athletic fields and the courses of endurance in Jacks Valley, I enjoyed the academic challenges. I was not afraid to answer or ask questions. I had been in a foreign academic environment in Costa Rica when all my classmates were speaking a language I could not understand. I was thrilled for the chance to ignore height and strength and gender and vie for academic equality. I do remember feeling so much freer when I stepped through the doors of Fairchild Hall, the academic building.

The academy promoted the "whole man concept," expecting to train and expose cadets to varied challenges. This complete package idea is attributed to General McDermott, the academy's dean from August 1956 through July 1968. McDermott's "biggest contribution was trying to avoid the rigid approach to science and engineering course requirements that had long been enforced at rival service academies." General McDermott introduced close to 30 academic majors to the Air Force Academy with some options in the individual curriculum.[10] This was the whole man concept. Cadets joking called it the "manhole concept." I didn't mind the gender reference. "Whole person concept" worked, but it wasn't as much fun to mock the requirements with "person hole concept." The positive effect of the whole man idea was that no one, or very few, were good at everything, but there were military, academic, athletic or emotional and even artistic undertakings offered, and often required, of all cadets. My inability to battle well with the pugil stick did not preclude me from succeeding at USAFA. There was plenty of challenge to meet the gamut of ability, experience and diversity.

The format of the days changed to accommodate classes and homework. We still rose with the sun to form up and march to breakfast. We scurried back to our rooms to grab books and head to classes until the noon meal march. We ran back to class again in the afternoon. Every day after classes, we participated in marching drill practice or intramural sports. We marched for the third time that day to the evening meal and were released back to the rooms again for a training time called "military call to quarters." The days closed with "academic call to quarters" for studying and finally taps, the bugle notes sounding from the speakers in the hall at 2300 hours.

We still kept our room inspection neat and made our beds according to regulations using a gray wool blanket folded down exactly 12 inches

Doolies in a classroom, lots of books, and lots of credit hours each semester. Back to front: Unknown person, Sandy Darula, unknown person. Courtesy of Clark Special Collections, McDermott Library, U.S. Air Force Academy

and stretched taut over the sheet. It was against the same regulations to sleep during the day, but after marching and mentally exhausting classes of aeronautics and astronautics and philosophy, the same blanket pulled me in whenever I returned to my room. I knew of many cadets who would not return to their rooms during the day for the very same reason. I think my blanket attraction was stronger or my resolve weaker because even if I had a short time before the next mandatory function, I would dash back to my room, drop my books on my desk and slide onto the blanket. We all called these covers gray magnets.

In addition to all the academic knowledge, the postures and routines for drill and the plays for intramurals, we continued to learn knowledge from *Contrails*. It was a new vocabulary to assimilate like total immersion into a foreign language. I thought I might have some advantage because my father was career military, a retired lieutenant colonel of the United States Army. But he retired so early in my life; I hardly knew more than a smattering of military words. When I was eight and he was still in uniform, he sometimes yelled, "Fall in" when he came home from the post in the evening. All his offspring shot to some child's version of attention: our backs arched, our bellies jutting out and our arms swung way back toward the walls behind us. We did a better impression of a penguin than of anything resembling a true military attention. The creases still knife sharp in his uniform khakis, he marched in front of us for a mock inspection. Waiting for his next command, we giggled and tottered. At "fall out," we all fell to the floor in a fit of laughter. I never saw it happen at USAFA, but at least I was familiar with those commands. Some knew nothing at all. Others, steeped in the military life from birth to their entrance in June 1976, knew much of it.

Karen Wilhelm was prior enlisted having joined the Air Force after her high school graduation in 1974. She intended to wear Air Force blue long enough to be eligible for the GI bill and get help funding college, but she rethought her plan as she worked on B-52s and KC-135s on Anderson Air Force Base in Guam.[11]

"Overall, being prior enlisted was a huge advantage," she remembers. "I knew how to wear a uniform, polish shoes and boots, march, salute, say 'yes, sir—no, sir,' run a buffer, make a bed with square corners, and all those other little details of military culture." She was also older with "real" Air Force experience, so she felt "insulated from some of the worst abuse. They knew that I knew there was a line they shouldn't cross." Yet she was still surprised at the stressful "leadership by yelling" approach, of upperclassmen getting in her face. She couldn't identify the value of anyone treating subordinates like that. "And don't get me started on eating meals at attention!"[12]

Forty percent of the women were not prior military themselves but military daughters.[13] Susie Park's father was an active-duty colonel in the Air Force. I jokingly wanted to call for some type of unfair advantage penalty against her.

Somehow, though, advantages always seemed offset by lack of information in some other arena. Sometimes it was current events or cars or football. It all depended on the cadet in charge and his emphasis of the day. Basically, Doolies were not allowed to know too much of anything. If an upperclassman asked you anything—ranks, aircraft identification—and you knew the correct response, or worse, if they sensed any smugness or pride, they switched to another topic until they found one in which you could be bested. There were few opportunities for any Doolie to have any advantage. It might be considered arrogance or cockiness and would be squashed like a bug on a jet fighter canopy at Mach 1.

During the day, any cadet might ask any Doolie for the menu for the day's meals or might throw out a question about academy history. With our chins in, at attention, we were called a "Doolie" and then harangued about the meaning of the word. If, with a serious military tone and in stiff military posture, we could spew out, "Sir, a Doolie is that insignificant whose rank is measured in negative units; one whose potential for learning is unlimited; one who will graduate in some time approaching infinity," then we would be correct according to the definition in the back of our little blue bibles.

In the evenings, though, it was hard-core training time with the Class of '79. During military call to quarters, these Two Degrees were primarily responsible for bringing the Class of '80 up to speed on all the required *Contrails* knowledge.

They often requested the "discipline quote."

"Give me the discipline quote, Utley."

"Yes, sir! The discipline quote from the speech by General John M. Schofield to the graduating class from West Point 1879," taking a breath and standing in the most erect posture. *Get this right and they will move on to someone else.*

"Sir, 'the discipline which makes the soldiers of a free country reliable in battle is not to be gained by harsh or tyrannical treatment.'"

"Why are you whispering, Utley? When I say I want to hear the discipline quote, I want to hear it loudly and proudly, not from some little mealy-mouthed Doolie who knows jack about discipline or the military. I can't believe they even let you in here, Utley. What are you doing standing there looking at me, Utley? I told you to recite the discipline quote and we are going to stand here all night if that's what it takes for you to get it right."

"Yes, sir. 'The discipline which makes...'"
"Utley?"
"Yes, sir?"
"Didn't you forget the introduction?"

Technically, I had not. I had already said it once. But without blinking, to hopefully save myself from another lecture, I replied, "Yes, sir" and "No excuse, sir," thinking that I could give my no excuse rationale in advance and save myself the litany of back-and-forth with the five responses.

He was distracted anyway by a classmate walking by rolling his eyes in upper-class sympathy to the horrors of having to stand training any Doolie, let alone a female SMACK. I hurried right ahead hoping he would not notice. My thoughts broke up the quote to relieve the tedium and to let my pride covertly win a point.

"Sir, 'The discipline which makes the soldiers of a free country reliable in battle is not to be gained by harsh or tyrannical treatment [*as if anyone in this whole crazy place ever even read those lines let alone processed the meaning*] on the contrary [*I wanted to scream*] such treatment is far [*far, far, far*] more likely to destroy than to make an army. It is possible to impart instruction and give command in such a manner and such a tone of voice as to inspire in the soldier an intense desire to obey. The one mode or the other of dealing with subordinates springs from a corresponding spirit in the breast of the commander [*as if they even have hearts within their chests*]. He who feels the respect which is due to others cannot fail to inspire in them respect for himself while he who feels, and hence manifests disrespect towards others, especially his subordinates [*hey, you, are you listening? This is the part that pertains to you*], cannot fail to inspire hatred against himself.'"

"Dismissed, Utley."

We memorized *Contrails* in sections. There was specific "knowledge of the week" which upperclassmen then dragged out of you while you stood ramrod erect out in the alcove by your room during "military call to quarters." The name was somewhat deceptive since had we truly been called to our quarters, we would have been *inside* our rooms, hopefully with the doors shut, our shoes off and our shoulders not back or down. Words had new meanings and phrases were confusing. Gail was reprimanded for wearing her hat indoors. Her "cover" which was her hat but in civilian speak, under cover meaning inside. Salute while under cover, salute when you are wearing a hat, but no, no cover while under cover. Don't wear your hat inside. Aaarrgghh.

Nevertheless, by now, in accordance with our training, we didn't try to draw logical conclusions from instructions. We just followed them.

Doolies sat in their rooms, uniforms completely immaculate, waiting for a Three Degree to test their knowledge. In some rare cases, it would be a Two Degree or Firsties, either of which would only demean himself to test Doolie knowledge with the most bitter of attitudes. If their car hadn't just been wrecked during a weekend downtown partying, or if their girlfriend hadn't just broken up with them, or if they weren't restricted because of marginal grades, they had better things to do than to stand six inches from some stupid SMACK's face and ask them to recite from *Contrails*. There we sat, the fourth-class Doolies, at the ready, hoping that the assigned Three Degree had forgotten, fallen down the stairwell while coming in from the evening meal or lay on the floor of Mitchell Hall having been run over by Mr. Martinez and his large steel hot-food cart. But inevitably, he came. He yelled out, "Hey, mister (or miss), get your ugly self out here," and we Doolies recited, "Sir, the discipline which makes…." That was military call to quarters. It happened to all Doolies, every night Sunday through Thursday … ad infinitum.

But again, the physical separation worked against the women. In this we were not Niners or Doolies the way the men were. We could not hang out in our rooms waiting for training because it all took place in the squadron areas, not where the women lived. Why would we choose to march into the lion's den? Instead of considering the pros and cons of the situation, we went with the easiest solution: the truth. And the truth was that the ramifications of not going into the men's area brought rumors of disparity and favoritism.

We straightened our uniforms after dinner and headed down to our squadrons to find a haven until training began. Some evenings, we hung out in one of the common rooms alone, but that didn't help our integration cause. We asked male Doolies to let us loiter in their rooms. Because of regulations, though, we were not allowed to close doors if opposite sex cadets were inside a room together. With women in their rooms, with the door opened to the upperclassmen's view, our classmates lost their anonymity and privacy. But there was no other way.

Our classmates were not the problem. The men of the Class of '80 mostly accepted us. They knew our main struggle was to fit in. Plenty of them held no grudges. They had never been a part of a single-sex USAFA. They knew no difference. Fred recounts that he was stopped on the terrazzo and harassed for having women in his class. It was not his decision to allow women into the academy, but he was part of that time in history. He and all our male classmates had to live with the evolutionary chaos that occurred at USAFA because of the acceptance of women. So many were our champions. Most were kind and considerate and encouraged us.

Marianne tells the story of a Doolie classmate who taught her an essential skill: "I hated to run," she said. "I thought running was my coach punishing me and making me do laps around a field." Classmate Mark H. encouraged her. "I am going to take you out and show you how to run," he offered. He recommended she think of it as being lifted up instead of being pushed down. "Think of a string attached to your head and pulling you up to the sky," he suggested. "For a guy to do that and show me how to take something in stride that I absolutely hated was wonderful. It was absolutely wonderful," Marianne cheerfully adds.[14]

Harry Lalusis, a squadron mate of mine that first year, chided us that we female cadets loved him because he could shine shoes. He could and we did. He could spit shine combat boots or daily-wear black low quarters until you could look down and check your hair in their luster. He was a master. He helped us with the shining, but he also taught us the process so we could always sport shined shoes that reflected light as if they were patent leather.

"Get a thin white cloth wrapped tightly around two fingers to start," he told us. "Tight, really tight so the fingertips pounded with your heartbeat. No wrinkles. Then spit into the can of polish and dab up the tiniest mote of black wax."

"Do we have to spit?" I asked.

He rolled his eyes. "Yes, spit! Gobs of it."

I used drops of hot water from the sink if it was available. I was not an accurate spitter.

"Then draw tiny little circles like mini Spirograph designs laying down a thin layer of shine at a time," he instructed us.

Roommates and I shined and talked until the backs of our arms ached, the shoes glistened and we got to know each other.

We really loved Harry because he cared about us. He did not just walk past us. He stopped to help and to genuinely encourage us. Harry's, like Peggy's, journey to the academy began with a dream to be an astronaut. "As a kid I avidly followed the space programs and dreamed of going into space. After reading Astronaut Michael Collins' book, *Carrying the Fire*, I knew step one was to attend the Academy and become a fighter pilot."[15]

He had attended the academy prep school not for academic enrichment but because his congressman submitted two candidates to the academy for consideration with only one slot available. Harry got a slot to the prep school as a kind of a runner-up prize. It was, though, a guaranteed place holder for Harry to join the Class of '80 and our fortune to have him then for our classmate. He had already learned how to shine shoes with secondary skills in sock rolling and skivvies folding.[16] Thanks to God for all the Harry Lalusises. There were many. They had faith in us, sometimes even when we didn't have faith in ourselves.

7. Doolie

Together with these men of our class, we grew accustomed to the knowledge questions and the military call to quarters routine. More rites of passage. We all did it, men and women. It was part of the initiation routine under the heading of Doolie year. Like shining shoes or calling minutes out in the halls before meals, this knowledge was part and parcel of the requirements of being a Doolie.

Living separately was the real problem.

Since our first day in the cadet area, all the women were billeted together on the sixth floor. This decision was not arrived at without discussion or disagreement, but I did not know that at the time. One of the options for the women was a separate squadron altogether.[17] It would have been a 41st Squadron with no history, no upper class and no men. That would have been a colossal mistake. We would have been an add-on appendage to the entire wing, no integration. Thank goodness that option lost its backers.

The planning committee decided on integrating only 20 of the 40 squadrons at first, but not all agreed. Some wanted to integrate the whole wing at once so as not to create a "split between the haves and the have-nots."[18] These doubters proved prophetic.

During BCT, especially at the beginning, we didn't know what difference it would make. The upperclassmen corrected us. The ATOs corrected us. The ATOs banged on our doors in the morning to startle us from our beds. The upperclassmen were waiting in the halls to continue the startling awakenings. Novices to cadet life and academy ways, we knew nothing else. We had no comparison. We didn't know what happened in the men's areas. As we marched down the marble strips on the terrazzo toward the dorms in the evenings, some male classmates peeled off at the column of every stairwell to enter their squadron areas leaving the strips dotted with the women all heading to the last stop on the line: the stairwell to the women's area.

Often, I liked being away from the same upperclassmen who demanded knowledge from *Contrails* whenever you passed them in the halls. I did not miss being so readily available to our cadet trainers. *Let them walk over here and climb the stairs to find me for their tirades.* Sometimes they did.

Up in the women's area disparagingly called the Penthouse, Shady Rest or the Ivory Tower, classmate Kathy Conley witnessed "special training." Overzealous upperclassmen came to the women's area to give "extra training" to the women cadets as overcorrection for the perception that the women were escaping the required training of USAFA. "Miss, do you think you're special?" or "Are you batting your eyelashes at me, SMACK?" they would yell. Today, it would clearly be sexual harassment, but we didn't have the vocabulary for it then.

The ATOs were more lenient in the training. They asked or told what was required and let us walk by, accepting only our "Good evening, ma'am" many other times.

But as the harsh initial training gave way to more mentoring and explaining of academy lore, we realized the disadvantages of living separately. The male cadets, for example, knew how to stretch a brand-new USAFA T-shirt over stiff cardboard cut to exactly 6 × 9 inches. Pinning the T-shirt tightly around the cardboard, they made a neat, false cover for the messy pile of shirts in their drawer. Those cover shirts stayed pinned to the cardboard for four years hiding the others that were actually worn. They were the hallowed shirts that many cadets unpinned and wore finally on graduation day. That was not in any regulation book. It wasn't written anywhere, but the upperclassmen knew it. So did our male counterparts who lived among the upperclassmen. We learned it too, eventually, but not from the ATOs who had never really been cadets and never worried about a four-year plan for keeping the drawers inspection-ready and neat.

The upperclassmen told stories passed down from the historical "brown shoe" days when cadets wore brown shoes instead of the regulation black. "Back when men were men and so were cadets," they'd start. Although that opening always made my jaw tighten and my eyes narrow in anger, I knew the subsequent stories were parts of the academy's history. Wearing the blue uniform, even if it was a skirt instead of pants, I had earned the right to hear the stories and to tell in the future, without the "men were men and so were cadets" qualifier. The ATOs could not tell the stories except as hearsay, inadmissible in cadet circles.

Living in the same hallways, the upperclassmen got to know the male Doolies. The Doolie men also got to see the upperclassmen differently. We missed that opportunity. We rarely had chances to see the upperclassmen as humans. When they made the effort to come to the women's area, they were training us, picking us up for a mandatory meeting or checking on the status of our rooms. We didn't meet them casually in the halls when they might distractedly just say, "Hi, Utley. How's it going?" We knew them on a very professional basis, appropriately. But with those few but more casual off hours spent separated from the squadrons, we missed the opportunity to know them as people and for them to know us when we were not stiff at attention or overwrought with anxiety. I never overheard them talking to each other as they passed me in the halls. I missed walking back up to the squadron with them after an intramural win and have the chance to talk about my family or theirs. We had to take the stairs while the upperclassmen took the elevators, but the possibility of talking while walking did not exist for me or for any of the female cadets.

The separation fostered rumors too that grew like a mold in a dark, damp basement. The stories of what happened in the women's area were wild fantasies that, like the whispering telephone game, grew crazier and more distorted with each telling. Classmates were discouraged from being in our area. We generally went to them. They did not come to us. Stopped by the ATO at the desk at the end of the hall—the sentinel guarding the fortress—our own squadron mates were questioned about why they were in the women's area. They avoided the area and started to question our training as well. The daring would ask us about the supposed TVs in our rooms or whether or not we had to make our beds in the morning. "Are you guys crazy?" I'd answer them. "Do you think they are trying to keep us? Why would you believe that they are trying to make things easier for the women that they never wanted in here in the first place?" But the rumors grew until the whole area became a glorious "penthouse" with room service and the ATOs as our maids who picked up after us. The word "penthouse" was even written in a published academy history to describe it.

The ATOs tried. They did try. Based on their simulated cadet experience, they offered us their best advice, but many things they just didn't know. They could not understand what they had not lived. They couldn't be what they were not. They did train to gather data for the plans for our arrival and eventually for our achievements, but their abbreviated course gave them only a taste for cadet life. Their Basic Training was only two weeks long. They did six weeks of Doolie life.[19] Plus, they were full-fledged officers. They outranked all the cadets who trained them. That must have distorted how they were treated. They would not have been called bitches or beavers or stupid. At least not to their faces.

Captain Galloway was part of the planning committees for the arrival of the ATOs and the female cadets. Her experience compared to ours highlights the disparity between the female officers and the female cadets. She had worked previously with General Vandenberg, after whom our dorm was named, and joined the academy staff in 1974. One written history calls her the human "guinea pig" as she decided to try everything the fourth-class cadets were supposed to do. She kept detailed records to use for preparing for the ATOs. "Yet in spite of stressful situations Galloway was determined to keep her cool and not break down and cry. She was forced to pull her rank to control unruly cadets."[20] I cried. Many of our group of ladies shed some tears. We had no rank to pull.

There sometimes seemed to be some competitiveness in the air between the cadet women and the ATOs. Twenty-two years old, they seemed ancient to me at 17. The rank they wore on their uniforms would cost me four more years of enduring and embracing the academy environment. Graduation and lieutenant bars barely registered on my horizon. But we all fell basically in the same age group. If only for a few years, they could have been the first women at the Air Force Academy. They could have had their pictures in the papers and on television in lieu of the shots of them in the pre-acceptance publications where they posed as stand-ins.

But when Lieutenant Walter, one of the first ATOs for the women, was asked to compare their survival experience with ours, she said there was no comparison. She recognized that ours was brutal, admitting that "the women of 80 were battered, that the guys used all kinds of sexual innuendos on them and they were called all kinds of names, ugly pigs and fat dogs. All stops were out."[21] She knew that they would never absorb all the animosity for us. So many of the battles to belong we fought for ourselves with our own strength and determination, with our own blisters and tears and torn muscles.

They oftentimes seemed to be our greatest impediment to integration and our biggest adversaries. To me, mostly they were symbols of our separateness and our distinction. Symbols that I didn't want.

Peggy, Debbie, June, Marianne, Gail and I would head down to the "Niner" squadron area regularly to check in and to have our faces seen. All I wanted was equal treatment. It frustrated me that during all this training, I would not achieve equality because living in this separate area tainted the value of my performance.

We all realized the detriment of this arrangement by now. We pined to start the school year like our male classmates, physically in our squadron areas.

8

NRA Shooter

> No one saves us but ourselves. No one can and no one may.
> We ourselves must walk the path.
> —Buddha

During that first year, I ended up on the pistol team. My interest in the team began with a stray comment from some upperclassmen, I think. We were not allowed to socialize with anyone of a higher level at all. That meant anything. No smiles. No hellos. No friendly talks. Nothing. It was basically a five-response type of relationship. Theoretically.

As a few Three Degrees from my squadron began actually treating me as a human, our talks got more personal just in the most basic way. "How was your day?" they might say as a passing comment. Just basic human connection, without training. One suggested that I "try out" for the manager for the pistol team. No skimpy uniform required. No dates or reciprocations promised. Just come down to the shooting range in the field house and talk to the coach about helping out.

It was fate. The team coach had asked them if they knew any personable freshmen willing to keep the stats and help with the equipment. I don't think he specifically asked for a female cadet. I guess I was being personable that day. Not really an athlete in high school, I had no idea what a manager position would involve. I certainly had no expectation I'd do any actual shooting. Many of the women were proven athletes from high school: all-state swimmers, number one in their state finals in track or multiple-lettered volleyball and gymnastic champions. I was none of these. I did letter in high school in tennis and volleyball. What no one on the academy admissions board knew was that I never won a match in all the tennis that I played, and I was the shortest player on the volleyball team. In fact, I was almost the only player on the team. There were really about seven of us, but some never showed up.

I asked about the manager's position and ended up on the full-fledged women's pistol team. It turned out that I could shoot. It surprised everyone. I had done some shooting with my dad in high school. He and I killed many an empty can or beer bottle out at the city dump in the hot Florida sun. I did it to spend time with him. I held the .22 caliber pistol in my hand, sighted along the barrel and pulled the trigger ever so gently on his instructions. It doesn't now seem at all like a thing I would be interested in or would be able to do well. Dad's comfort and calm with a gun in his hand came from growing up as a hunter. In his 26 years in the Army, he also learned not to be afraid of handling a weapon. He passed that composed assurance with a weapon on to me during the landfill target practice, our father-daughter time.

At the range at USAFA, I asked for a turn one day at practice, breathed slowly and deeply and left some light holes in the black bull's-eye of the target. From then on, I was a member of the Air Force Academy pistol team and usually one of the top shooters on the women's side.

With my team membership came an NRA card, shooting every afternoon instead of marching drill practice or squadron intramurals, and the privileges that befell an academy "jock." Privileges for athletes included team practice instead of squadron drill or intramurals every other afternoon and eating meals with your team and without the stiff posture and no-talking rule of the squadron tables. Those were not to be scoffed at.

* * *

Lisa Lambert ended up on the team too. She ran into a female classmate in the latrine one day and had a conversation that was not allowed in the bathrooms about how the pistol team was looking for women cadets to fill out a roster. The classmate let her know "how nice the upperclassmen were on the team and that there were team tables." That was perfect timing for Lisa. She would welcome team status over intramurals which were required for all non-team cadets. Her squadron flickerball team had just trampled her, then traded her to the squadron cross-country team. She did surprise herself by running six miles by the second practice. Cadet life did push us to perform beyond our own expectations. In a 28 September 1976, letter home, Lisa wrote, "They can use 10 girls [on the pistol team] and I made up the 5th one. They said the only way I would not make the team is if I shot the instructor."[1]

Karen Wilhelm appreciated the different environment that came from her participation in the Cadet Drum and Bugle Corps. Karen acknowledged that some of the upperclassmen in the corps were STRAC, an adjective borrowed from old army lingo for Strategic Army Corps that morphed to Skilled, Tough, Ready around the Clock and came to mean someone sharp and with their shit together. "Nobody dumped on you because the focus was on the music and learning the marching show,"

8. NRA Shooter

Karen continued. "We got to travel to all the away football games, marched in parades and did other events in the front range area, and had three big trips every spring to parades and music festivals around the country."[2]

Ostensibly to build team unity, all teams also sat together at meals, which we called "sitting on ramps." With the remainder of the Doolies still eating in a stiff square posture, sitting on ramps, "at rest" together with our team, was no small thing. Eating at rest allowed relaxing, eating and talking with others at the table. The atmosphere was a quantum leap from the normal squadron tables. No grilling with knowledge questions as you waited for the meal to officially begin. No sitting in a stiff braced position while you ate. Upperclassmen on ramps did not end the meal with the command "Post," which required Doolies on squadron tables to have a joke or quip to recite that had to be rehearsed ahead of time or secretly passed under the table to your classmates. On these athletic tables, we escaped all the terrors of the meals and actually ate something. That was one of the great appeals of the teams.

One group that was very obvious in their status as ramp eligible were the cheerleaders. All these years later, I realize how unfairly they were treated and how unfairly I judged them. I was enjoying the privileges of membership on a team that shot guns and seemed ultra-military. But during the day-to-day life along the marble strips and within the halls of the academic building, no one could single me out and recognize me with that privilege. The cheerleaders were very recognizable.

There had been no freshmen cheerleaders ever and no female cheerleaders ever. Our Doolie year, freshmen and female cheerleaders were introduced all at once. Many in the wing shunned and harassed the female cheerleaders because they got privileges before their traditional time.

Like any other change instituted our first year, it was attributed to the women's acceptance, written down under the heading "Problems That Marched in along with the Women," like a list of our mortal sins. "Your class is weak" was a common admonition because we were being trained with positive motivation in lieu of the "drop and give me 20" training formula of years past.

A woman doing anything in the cadet area our first year was observably a Doolie. On a dare, male classmates breached the air gardens, a small area of the terrazzo near the academic buildings that was forbidden to Doolies. Walking back from the library one dark night as a Doolie with male classmates, the men bee-lined it through the air gardens with their books covering the bold white 80 sewn on their parkas that would have betrayed their identity as freshmen. A quick glance back to me and they knew I couldn't do it. I did not have enough books in my whole room to cover the two nylon-covered legs protruding from my skirt that gave me away. We couldn't blend in. But the cheerleaders were particularly obvious because they stood in front of the entire wing at each football game.

I also struggled with the female cheerleaders because their difference was so obvious. They did not fit in the mold of the female cadets that I thought the rest of the women were trying to establish. I never felt the same about the men on the cheerleading squad.

The cheerleaders took extreme harassment from the cadet wing, and they were attributed many unearned adjectives. Flirty, but maybe it was just friendly. Maybe it was just the job description of the cheering squad. Maybe they were setting exactly the right example for cheerleaders. But they seemed so unmilitary because they wore the uniforms of the cheer squad, short white bouncy skirts with Air Force blue pleats and a cute blue sweater with the Air Force Falcons lettering. That was a uniform that let them look like real women. The cheerleaders could stand next to the cheerleaders from Wyoming or Colorado State and look normal. That was a privilege the other women did not have. Why didn't they cheer with their rifles slung over their shoulders in their full fatigue uniforms? Was I jealous of their obvious femininity, or was I acting like the unyielding men of the upper classes wanting all the women to fit some outdated historical and military mold? I somehow wanted the cheerleaders to be the cadets as defined years before even though none of us were. We were all changing that definition with our own talents and interests. At the Falcon football games, the cheering women took much taunting and boos accompanied by thrown oranges or paper cups. I never threw anything, but I am ashamed that I didn't stop anyone from throwing anything.

I have to admit, too, that team meals seemed like an unfair advantage in my own heart. All my struggling to be an equal with the men of my class quieted to a nervous tic as I sat on ramps. I wonder if the academy tried to incorporate the women as soon as possible into athletic teams to show our complete acceptance. Intercollegiate sports teams were formed from the 157 women available, giving slots to those that in future years, with more women to choose from, might not have made the cut. There were women all-American swimmers, true track stars, serious basketball players who certainly earned a position on intercollegiate teams. But all the female teams were formed in that first year, all with freshmen, Doolie members. Because the pool of participants was small, many got jock status and sat on ramps, at rest.

I quieted the noise in my conscience with the knowledge that I could actually shoot. I was one of the top female marksmen at most meets. But would I have been chosen from among 600 women who would be cadets in three years' time when all four classes were coed? I swallowed my guilt as

8. NRA Shooter

I swallowed my food in Mitchell Hall without my throat restricting every bite like it would have if I had been sitting at attention.

The break at meals and interacting with teammates in the field house gave me relief for the time that the team was competing. I still got upset with the other Doolie training that demanded so much of my emotional stability. I took some of my frustrations out on the targets. The back of one became writing paper sent to my brother with a small, tight grouping of holes just to the left of the bull's-eye. I think the gun needed a slight sight adjustment, but otherwise I think I was dead-on....

> Pat,
>
> I'm writing first because I owe you a letter and second because I need to talk to someone. I just had a long talk with my roommate, Susie, who has become almost like a sister to me, and she told me she is leaving!! I couldn't believe it. She seemed to be one of the most motivated girls here. It is hard for me to understand why, yet in some ways she is right. The system here stinks! The girls are completely separated from the rest of the wing. It's like we belong to a separate Academy. I keep getting my hopes up thinking about moving down to the Squadron area with the guys, but most people think it will never happen. It will never work like this. I am not leaving, but I won't stand to be second rate. I'm not a women's libber and you know it, but it is not fair for us to be separated from everybody else. Once we get into the Air Force, we are all going to be together. I have been crying all afternoon about it.
>
> I am not going to leave. I am determined to prove that this system will work. They are definitely going to have to move us down into our squadrons, but it will work. If anyone can do it, I can. I'll show them. I know this isn't exactly the happiest letter I've ever written but it has been a bad afternoon. I feel better now, thanks to you. I will write again when things are a little brighter....
>
> Bean

The words were so unlike the paper; he must have chuckled as he read it. On the front, target side, were a tight grouping of almost bull's-eyes. An official ten-meter target of the NRA, it measured about seven inches square. Ten concentric circles radiated from the center, numbers ten through seven inked in solid black for the bull's-eye and the remaining six only outlined in black. The five bullet holes shot from the prescribed 50 feet away by my .22 caliber pistol pierced the black section only. The tallied score was 41 out of a possible 50 points. The tight grouping showed accuracy and consistency in the shooting. With an adjustment of the weapon's sight, that same tight circle could have easily been all nines or tens.

One would not think that the coldhearted, unemotional shooter sat at her desk writing woes of her sad story to her big brother. The emotions of the letter, written circularly through the tan circles on the front and continuing over to the back missing the punched holes, just did not match the steady, deadeye impassivity reflected by the bullet holes. I was determined to make it at the academy but also lamented and whined about a roommate who was leaving. Why? USAFA was a tough place. People left. Not everyone was cut out for the 0530 wake up, the morning runs, the 21 credit hours per semester or all the yelling and criticism. No one truly was prepared for that. The NRA card-carrying shooter who blasted that target should have been able to handle a roommate leaving the way she obviously handled the gun, without flinching. Be tough and gut this one out and get on with making your place at the academy.

But it wasn't that straightforward.

There were two major issues in the writing on the target. One, the separation was wearing on the women cadets, on me. We were exhausted trying to belong somewhere when we weren't being let in to have the opportunity to belong. We needed to be physically in the squadrons to be able to be assessed in our abilities against the traditional successes of male cadets and in parity with the men in our class. This was an ongoing academy problem.

Secondarily and very personally for me, a roommate was giving up. That had a massive impact because all cadets, men and women, got emotionally attached to our roommates, our cellmates, our confidants. The system was built that way. Except for the pugil stick competition, every endeavor, every challenge so far was predicated on the concept of teamwork. There was little expectation of excelling or failing alone. Academics was also more independent, but squadron mates and classmates even helped each other study and assisted with projects as far as the honor code's "thou shall not cheat" clause allowed.

Our roommates were the classmates who "checked us off" as we left the haven of the room. On them depended the status of our uniform, mostly the parts that we could not see ourselves. The condition of the back of your uniform was wholly the responsibility of your roommate. Upperclassmen held roommates accountable if his or her roommate appeared on the terrazzo looking anything but immaculate.

A roommate was also a coconspirator who might share mailed contraband from their grandmother's kitchen or gum from their dad's care package. The roommate helped with homework and room work, the cleaning, the polishing, the positioning for inspections. Once when prepping for a room "white glove inspection," Peggy and Gail were polishing the mirror and vanity and buffing the floor while I lay in the bed sick with a stomach bug. We would all stand at attention for the inspectors in the morning, but I felt queasy and weak when I tried to get up to help them. They didn't flinch. Roommates carried each other when needed. They did squirm a bit when I threw up in the sink right after they had scrubbed it, but they cleaned it again and told me to rest.

Classmate Bonnie Houchen had similar support and not just from

8. NRA Shooter

roommates. She was excused from a morning run with rifles but not, she learned with no notice, from having her room ready for inspection while her roommates ran. Upperclassmen told her she had ten minutes to prepare the whole room. Immediately, female classmates from all different squadrons came in ready to help with cleaning supplies in hand. Bonnie says, "They told me to get dressed and they would take care of the room. The look of surprise on the upperclassmen's faces was 'priceless.' They could not believe the room was ready for inspection," Bonnie recalled. "It would not have happened without the help of my 'sisters.'"[3]

Roommates were the stand-ins for the sister or brother, the best friend, the parents who could not be there or even call for those first few months. We connected with our roommates beyond what the walls of our room defined. You knew your roommate's hometown and high school sweetheart's name. You knew her brother's favorite sport or all the stories about her sister's antics. You knew their weaknesses—academic and emotional. You built your family in their mind so they could conjure them up with a story at any time without involved explanations. Good roommate relations were essential to USAFA success.

Susie Park was my second roommate. Not a Basic with me in Invaders, Susie had been a Basic cadet in Hellcats and now for the remainder of her Doolie year, she belonged in Eagle 8th Squadron. After BCT, because of constraints of the living spaces, we bunked at least two and sometimes three to a room. Ninth Squadron had six women and 8th Squadron had seven, so odd women would sometimes have to room across squadrons. After acceptance, roommates were shuffled. Peggy moved in with other Niner women, and I was now with this unknown, Susie. She was the remainder after the rest of the 8th Squadron women were paired up. I had no problem with Susie at all. I loved her like a sister, more than my own sisters at times because my dependency on my sisters had not been tested so often or so harshly. But having two squadrons represented in the same room doubled the trouble more than it doubled the fun. When upperclassmen ventured up to inspect the women's rooms, we might get two inspectors, one from 9th and one from 8th. Inspectors from each squadron intruded to run white gloves in their favorite dust-hiding spot, under the rail of the bed for one and maybe up inside the track for the sliding closet doors for another. Unknown, unfamiliar upperclassmen came by to ask knowledge questions. Allegiances for squadron cheers and intramural activities got separated as if by childish masking tape, dividing the room. The men never faced that confusion. As far as I know, they never had to room across squadron lines. No matter how odd the numbers, squadrons made pairings or triplings to keep the squadron men within its physical boundaries. Niner men all lived with Niner men.

But Susie was fun and worth the extra hassle. When I asked her why she thought to apply to the academy, she told me, "Dad always wanted me to do it. He stayed on top of it." He was still active duty, so she, like me, like many of the women I think, got a presidential nomination. Her friend also wanted to go, so they did their testing together. Her friend couldn't clear her ears because of some eardrum problem, so was medically disqualified. Susie was going to be denied medically for acne and a previously torn ACL. Susie registered at Louisiana State University for their pre-vet program while her dad worked toward and got waivers for her for both conditions. She told me she saw *The Omen* the night before inprocessing. "What omen?" I asked. "The *Omen*, the 1976 horror movie: *The Omen*. I think it was an omen!"[4]

She was so proud to be at the academy. She was enthusiastic like Peggy, loving the challenge and the adventure. She, like Gail, added defiance to the atmosphere. That defiance ran in the face of some of the hair regulations.

Susie had beautiful, long, advertisement-worthy hair in her high school senior picture. She remembers during inprocessing on day one when that hair disappeared with everyone else's remnants into the garbage bins in the barbershop. She said she "cried inside when she got her haircut even though she had already cut it short."

She seemed to lose confidence when she looked in the vanity mirror and saw someone else. Besides the chopped hair, many women gained weight. The male cadet lingo came to include Colorado hip disease, CHD, as an acronym for the women who had added pounds. All women were weighed in regularly, and the men upper-class cadets did the weighing. Susie explained, "We were required to go to meals. If you weren't eating you got ragged on, so I ate and got fat." The academy instituted salad tables. Some cadets, mostly women, went there to eat to escape the harassment at regular tables, but some were assigned those tables to force the dietary restriction. Susie kept that high school senior picture in a clear outside pocket of her ID wallet because it reminded her that she wasn't as ugly as she felt.[5]

Susie rebelled in her "beat the hair regulation" routine. After Basic Training, the regulations only required that female cadets' hair not touch the top of the combination Alpha suit coat collar. The regulation did not allow the use of any bobby pins or clips or ponytail holders in the hair. At first, she kept getting her hair cut as I did. At Thanksgiving, she flew home to Louisiana and her mom gave her a perm to spring the hair up shorter. Later in our Doolie year, Susie religiously washed her hair at night before taps. With tiny pink foam curlers, she tightly wound small sections of the hair, squeezing the foam tight with hair and clipping the pink plastic arm over the roll for overnight drying. In the morning, the release revealed curls like tiny springs that pulled up against her skull and away from the Alpha collar. I'm sure she kept her chin pulled in extra tight and shoulders down to tilt the hair upward even more. No regulations defined our civilian dress, so once we received privileges to wear civilian clothes again, or maybe when we wore them against regulations without privileges when we went downtown, Susie was brushing hair down almost to her shoulders. She glowed with her full mane released, but it might have filled many upperclassmen with rage if they saw her proud locks of defiance.

She also still struggled with acne. The dermatologist she saw through the academy medical system told her she would never fly and gave her no resources and no comfort. Susie's mom went to their hometown doctor and sent birth control pills from home as her acne treatment.[6]

Her father was an active-duty colonel, chief of security police for 8th Air Force in Louisiana, so Susie also added accurate military knowledge to the mix of our room.[7] She knew ranks, bases, and many airplanes on sight. She taught me hints for distinguishing the C-130 from the C-141. She could announce the ranks of officers with a brief glace at their insignia, leaving me to just mimic the titles she so easily attached. My father retired when I was eight. Susie lived her entire life as an offspring of an officer. That history left her less frightened as officers approached, more willing to ask for help.

She was crippled, though, with her knowledge of the "real Air Force." She agonized, as we all did, over the inequality that living divided from our squadrons created. She had seen the Air Force of our end goal in action and knew that we would all, male and female, work side by side, possibly in the same cockpit when on active duty. Her life was seasoned with the stories of the men and women of the Air Force. It tortured her to know that the academy was backward with our segregated living placement. There were no pink curlers to rebel against that separation although she spent as much time as possible among her classmates in Eagle Eight.

She was disappointed with our treatment too. Once when walking over the "Bring Me Men..." ramp, the ramp that was our portal into the institution, a passing upperclassman told her to "take a flying leap." She grappled to reconcile that behavior with her dreams of an Air Force career.

She struggled. She wanted to be a pilot but didn't think she would be permitted. She thought she could go into medical school or veterinary school but was told that continued professional education after USAFA was being discontinued. In fact, it was not discontinued, making me wonder if the speaker of that information was trying to get her, get any of the women, to quit. Susie's father came out to USAFA for a visit hoping to convince her to stay. She decided to try to stick it out to the end of our Doolie year.

Susie made her decision after academic finals and announced that she was leaving. It was not an easy choice. Her delight in her acceptance and her year of endurance along with the heavy pride of her father held her at USAFA like an anchor. The father weight got heavier when she got a letter from her dad saying they were building a house to retire in. "I am so glad you decided to stay," he wrote, "otherwise I wouldn't be able to afford to build this house." When she left, she reflected, "It broke my daddy's heart. I think at some point he forgave me, but he couldn't say it."

My mom even called her and tried to convince her that she could make it if she would just take it a day, a minute at a time. "I'm not a quitter," she told me years later. "I just know when something is not good for me." Quitting, as according to cadet speak no one "leaves" or "transfers out" of the Air Force Academy, was the topic of many nighttime conversations between Susie and me. The private exchanges provoked many misconceptions from others who walked in on our abrupt sentence stops and quick tear drying. One female cadet later confessed she thought us stuck-up and scheming against her, accentuating the paranoia and instability that we all lived with.

Sometimes I stepped away from Susie's tears and troubles. I had to keep myself a bit emotionally detached from the discussions. The thought of her leaving and having all my emotional investment and stability ripped away could only lead to another episode of my doubting my own rationale for staying. *What am I doing here?* I felt on such shaky ground in my own decision to stay that I wouldn't dare let myself think about outside possibilities. Shaky ground could easily become an earthquake. I listened and consoled but was secretly glad that I had not reached the point that I had to seriously think about quitting. For me, it seemed like it might truly be harder to go than to stay.

Susie became a pariah. As required, as soon as any cadet decided to quit, he or she was assigned to an otherwise unused designation, R Squadron. The cadre members of this reject or rebel group wore different colored ascots with their uniforms as a sign of the quarantine the cadre enforced. For the quitters, the R Squadron routine included wearing low quarters, black oxford-type dress shoes, with the green fatigues, the utility uniform. Like high heels with overalls or wingtips with sweatpants, the two did not match up. It was a uniform combination that would never be allowable in any other situation. In effect, this was a neon beacon to all within eyesight that this low-quarter-and-fatigue-wearing cadet had given up, did not have what it took and was leaving USAFA, punching out. Let this be a sign to you all. Keep your distance and beware of contamination.

Was the thought that those who resigned had also given up on the honor code? Like a corporate executive walked out of the building, the R

Squadron cadets were escorted everywhere as if they would run around lying, cheating and stealing at every opportunity if not closely watched.

After her decision, Susie could have practically nothing to do with the rest of the cadet wing. She still did sleep in the room with me. I know we talked for hours into the night about what she would do and where she would go. She later applied and was accepted on her first try to LSU vet school and became a veterinarian with a successful career. She didn't regret her year at USAFA, happy with the friendships she made and what she learned about herself. For one thing, "she learned she didn't want to be told what to do by morons."[8] After signing her resignation papers, she did not attend any functions or formations. She said she thought about staying through Hell Week to finally reach the official recognition that would make us upperclassmen, but one upper-class cadet suggested she not stay for the torment but just leave. Another upperclassman, one who had admitted hating having women at the academy and who made constant comments to Susie about that hatred, made one last comment as she was outprocessing. "You know, Park," he commented, "I have a bunch of sisters. Can you imagine if one of my sisters had come to the Academy?" Susie, proud to have the perfect thing to say and then actually say it, responded, "If you could have imagined one of your sisters here, you would have been a lot different."[9]

The escorted outprocessing took only a few days. Then all traces of Susie were erased from the cadet area except a high school yearbook photo of a smiling confident young face, framed with long, beautiful brown hair, held to my bulletin board by a tiny silver thumbtack.

9

My Darling Daughter

> You are braver than you believe, stronger than you seem, and smarter than you think.
> —Christopher Robin, "Pooh's Grand Adventure: The Search for Christopher Robin"

In 1968, before Gail went there, previously all-women Vassar began accepting men. The early Vassar men spoke of the presiding feeling of loneliness during the first year or so. Even though being surrounded by women seemed perfect, many of the men found themselves craving "buddy" time. Vassar men began to pride themselves on being "brothers in a special sisterhood."[1] We then were sisters in a special brotherhood but in an unorganized fashion.

Sisterhood, though, was a feeble component of our struggle at USAFA. We had no women's support groups. We were trying not to be recognized as women. We were trying not to ask for support. I remember that we once were ordered to a meeting as a group for a talk by the flight surgeon about why we were all missing our periods. It turns out it was a stress-induced situation. We knew we were stressed. The meeting added stress, and I felt so uncomfortable to be in a forum of only women with a definite female topic of discussion. We didn't meet as a group of women cadets. It wasn't done. We were blending in, bonding with the men. I worried about how it would be perceived by the men that we got together for something beyond the normal training.

Although I don't remember it, there was also a mandatory meeting for a makeup demonstration by a Mary Kay representative. We concluded that maybe someone thought we were not feminine enough and needed some expert advice on making ourselves attractive. I thought we were getting enough attention. Femininity was a concern with the planners too. They often mentioned not wanting the female cadets to lose their femininity. Full-length mirrors in the latrines would not reflect our femininity. White go-go boots were not the answer. Female role models in the form of ATOs added a feminine presence to our experience, but there was no single description. Kathy Conley has a vivid memory of her march up the back stairwell to her room on the first day of Basic. The ATO escort, Lieutenant Paula Gathwright, told the group not to lose their femininity. Kathy was confused about losing femininity—what did that mean? The ATO explained, "It is something on the inside that can't be taken away."

Gail points out that within our group, no one was exactly the same. "No two women in my class were alike. We represented the full spectrum [of femininity]. No matter where we fell on the spectrum, we were critiqued by someone for 'why' we came to the academy, how we talked, our clothes, the pitch of our voice, our physique. For me, it was the curl of my natural hair."[2]

We ran the gamut from princesses with makeup and hair perfectly coiffed to unadorned tomboys comfortable in uniform fatigues and combat boots. I chose inconspicuous androgyny, although I kept a small vial of Charlie perfume in my medicine cabinet, sprayed mostly on weekends. Especially in my Doolie year, I would rather be considered drab and homely, one of the guys, than get any extra notice, good or bad. Perhaps that Mary Kay representative arrived in a pink car. Gail says we all received a sample lipstick in a "beautiful natural shade, bright pink." I'm sure I never wore any since it would have clashed with my orange hair. Gail didn't either as bright pink is not natural with her ebony skin. There was no defined cadet man or woman. There was no clear mold of femininity or masculinity. There was not one shade of lipstick for us all and makeup meetings didn't create some miraculous sisterhood bond.

We did benefit with support from home. What would I have done without the caring letters of my old-fashioned career Army dad? My mom wrote too. She filled typewritten pages with news from everyone. That's what Mom did. She was "communications central" by her own definition. She wrote once in response to one of my letters complaining about the Colorado cold and offered to send me some footed PJs. Thoughtful but wildly off the mark, though not long after, I could have used those PJs for a bunny costume. Mostly in her letters, Mom kept me filled in on the day-to-day happenings in the family, leaving my dad the privilege of just gushing pride and support. He knew what to write, and he knew the jargon. There were a few differences in the Army versus Air Force terms, but he needed no translations for NCO or AWOL. He understood what I was going through in experience and not just in empathy. *Non illegitemus carborendum* (NIC), his Latin philosophy, "don't let the bastards get you down," was part of each and every note. He knew it worked. He tested it and drew on it to survive his 26 years in the U.S. Army.

Dad's support was the bridge that carried me over some very rough waters that first year at USAFA. I ached and faded from physical exhaustion and from mental anguish after all the berating. But his letters would arrive to buoy me up.

A letter from Dad during Basic Training, 10 July 1976:

Dear Beanie,

How is my loving daughter? I love and miss you. I have a great big soft spot in my heart for my daughters, especially when they are away.

I put the old lousy Latin phrase on the back of one of your mother's letters and I want you to keep it in mind. (N.I.C.) You have always lived in a very serene house, where I never permitted any loud arguing or shouting, where Mother and I never did any fighting. So, the upshot of this background is that you may not be fully prepared for highly competitive strident shouting surroundings. Don't let it get you down. It is all a planned psychology. When I went to OCS [Officer Candidate School] all the candidates were sent out into the woods to learn to shout. We would shout at the top of our voice our name, rank and serial number, the Gettysburg address and anything else that came to mind, just to get used to talking and giving commands in a loud and attention-getting voice. So don't think there is anything personal at all—there is not. It is just an ancient philosophy of the military mind. Just tune your hearing down and do remember your Latin.

Honey, I love you and am pulling for you, and I know you can hack it. If things get to piling up on you, just haul back and think it over. That good think tank of yours will put things in proper perspective.

Love,
Dad

A wonderful treat during that tough summer, but I didn't think he was absolutely on target about things not being personal. It was so incredibly personal for us sometimes. Female cadets were such visible targets for the venom that the upperclassmen spewed. The issue was training, yes, but the serious, acidic words were directed specifically at us, the women of the Class of '80. No, they may not have been composed for the ears of Kathleen Utley specifically. In that respect, Dad was right. It wasn't personal to me.

When June's shoes were not shined or Debbie had a "gross cable" protruding from the pocket of her fatigues, that wasn't personal. When upperclassmen asked Gail to identify an airplane from a one-inch slide out in ranks at the noon meal formation one day, she could have looked around and seen upperclassmen holding similar slides up in front of some male classmate's face. She could hear some other upperclassmen's tirade, calling someone a stupid SMACK and asking why he didn't join the Army if he couldn't identify even the most common aircraft. That didn't feel personal. It was the common training for our class. Although it was tough, that was the kind of attack that most of us could let go. Well, not really let it go. For not knowing the aircraft by name, designation, manufacturer and nickname, upperclassmen assigned a 100-word essay about that aircraft due by 1900 hours that evening, squeezed in among three more classes and two hours of mandatory intramurals or marching drill practice—*then* you could let it go. There probably were three or four other squadron Doolies who had gotten the same assignment. So in those kind of cadet training situations, ones that everyone knew and experienced, it wasn't that specific to you.

But with only about 150 of us women there in a sea of more than 4,000 men, it felt acutely personal when the attack was about "bitches" or "beavers." When caricatures of women cadets with grossly exaggerated breasts or hips were taped up in the men's latrines or pictured in some underground cadet magazines that we all saw, it felt very personal. When a file of hundreds of Doolies walked down the marble strips after a meal rushing to get books for afternoon classes and the one female cadet in the hundred was called out, "You, bitch, get out here," we responded, "Yes, sir." We stood at attention through their venting about how things were better before women diluted the training. You knew that the men got something a little less personal than that. It was personal, directed at who we were as persons, female persons, women.

When I could no longer do it for myself, I did it for Dad. But I knew, too, if and when I had had enough, he would welcome me home with those strong loving arms. He would hold me up for another choice in my life if this regimented one was not in the cards for me. I would receive his love for making the tough choice to stay or the sometimes seemingly tougher choice to go.

July 25th, 1976

My Darling Beanie,

If I wrote as often as I think about you, you would have a big box of letters by this time.

I miss you a lot. You and I saw eyeball to eyeball on a lot of things and I could talk to you on so many subjects and you were always ready to gopher things (you don't know how nice it is to have another driver around) so I feel I have lost a good right hand with you gone. I am so happy for you and so proud of you that I think I bore a lot of fine people bragging about you. I can't even talk about you without sticking out my chest and feeling proud.

You are very highly motivated, I know, and it would take a lot to stop you. But, I want to tell you that you are not locked in. If you should ever decide that it is not for you, you are loved and wanted under all circumstances. We love you.

Fathers miss daughters very much, but I am so proud of you that it makes it bearable.

All our love and best wishes,
Dad

P.S. An old Grunt quote, "how come they never got your goat?—I never told anyone where I tied it."

The support from home barely offset the weight of the whole training process. I don't know how those without it survived. Some didn't surely, and others must have made up for it with their personal determination and drive. I am in awe of those. I cannot imagine how I would have made it without those letters from home.

An undated note from my older brother Michael, a Vietnam vet:

Beanie,

When you get this letter, you will be, no doubt thinking about quitting [as if we ever were not thinking about quitting]. I remember my first day in the army. This 6'3" Sarge came up to me and started bitching about the socks. [I guess Mike didn't have any on.] He said "Boy, where you from Boy?" "Miami, sergeant." "Don't they have socks in Miami?" "No, Sergeant."

Just remember that those people yelling at you are no better than you. Bean, it is a game that you're playing. All life is like that: just a game. Their rules are to get you mad and try to make as many as possible quit. Your rules are to obey them and realize what they are trying to do and come out above them.

Over in 'Nam a friend of mine was driving a jeep, and I was riding shotgun, which means that I protect, and he drives. Well, this hard ass officer came up behind us in his jeep. We had stopped because some planes were dropping bombs nearby and my friend and I wanted to watch them dive bomb their target. This Lt. Colonel was in a hurry, and he yells loud and fast "GET MOVING SOLDIER. Haven't you seen an air strike before?" My buddy Jim yelled back "NO, SIR." The Lt. Colonel stopped his jeep and jumped out. I thought he was going to kill poor Jim. He came up to our jeep, and said to Jim "WHAT THE HELL DID YOU SAY, SOLDIER?" Jim really said to him "No sir," meaning that he wasn't going to move for some asshole Lieutenant Colonel. But Jim said to him "No, sir, I've never seen an air strike before" which was a lie, but what could the officer do? He just said, "All right now get moving." We drove on down the road laughing at that clown.

Beanie, just keep your stuff together and you can beat them at their own game. But you have to be ready and always be thinking. Think of what they're trying to do. Just like in any game, if you know what their game plan is then about half the problem is licked. If you let whoever is yelling at you know that he is going to lose his voice before you lose your cool, then you will have no problem with the big dope.

Just remember all the thousands of men who have made it through the Academy, and you know that they have nothing more than you don't have, except a pair of balls and sometimes that makes them more vulnerable.

"You are strong, you are invincible, you are woman."

I love you,
Mike

* * *

My relief came from the Cadet Chapel too. This glorious building that was a hallmark for the academy and its plaza a perch for the gawking tourists was a peaceful haven for me on most mornings.

The chapel was designed by 34-year-old Walter A. Netsch, Jr., of Skidmore, Owings and Merrill Architects of Chicago. In the design phase for the academy, Air Force representatives stated, "We want the Academy to be a living embodiment of the modernity of flying and to represent in its architectural concepts the national character of the Academy.... We want our structures to be as efficient and as flexible in their design as the most modern projected aircraft."[3]

As tourists learn on the tour, "soaring 150 feet toward the Colorado sky, the Air Force Academy Chapel is an all-faith house of worship designed to meet the spiritual needs of cadets." It contains a separate chapel for Protestant, Catholic, Jewish and Buddhist religious faiths, plus two all-faiths worship rooms, "according to printed chapel information." The aluminum, glass and steel chapel has 17 spires. There supposedly is no significance to that number. I think the original design called for 21 spires to represent the 21st century, but cost constraints reduced the number.

When I was a cadet, I memorized the "fact" that the ultimate number, 17 spires, represent the 12 apostles and the 5 chiefs of staff. "Each spire is made up of 100 tetrahedrons. Each spire is 75 ft long and weigh 5 tons each. They were fabricated in Missouri and shipped by rail." The chapel received the American Institute of Architects' 25-year award in 1996, and the Cadet Chapel was on the cover of the April 14, 2007, edition of *Life* magazine which featured America's coolest churches.[4]

Cadets were required to attend worship services until 1973. Our class was not required, but I wanted to go. Going to Mass at on Sundays and at 0530 on most other days to find peace and particularly to escape the melee of the formations brought its trials, but it rewarded me with some occasions of calm. No surprise that the highest attendance for cadets is during their Basic Cadet Training.[5]

In order to take advantage of this cease-fire, as it were, I set an alarm and woke at 0500, instead of to the blaring of reveille that would come normally about half an hour later. I popped out of bed and got ready while my roommates still slept. They looked so peaceful and content, but I knew their tranquility would not last. I quietly headed out the door to many fewer people stirring for a semi-quiet run to the chapel. My path was always along the marble strips at the edges of the terrazzo even though at that early hour, few were outside to check. Once in the worship space, I sat completely "at ease," no talking but with a relaxed posture, and participated in the service. Most rules were disbanded within the chapel's beautiful interior.

The upstairs held the Protestant congregation area. There, stained glass windows filtered the light from outside. After the original design was considered too modernistic, the designer, Walter A. Netsch, went to Europe and drew inspiration from the Cathedral of Notre Dame and

Sainte-Chapelle in France for the beautiful mosaics of stained glass that brought light and slices of color into the sacred space.[6] The colorful aura transformed uniforms, floors, faces or whatever it landed on, no matter what the outside sky was like. The A-framed interior mimicked the outside of the chapel and brought your eyes toward heaven with the pointed inside of the spires.

Within the Catholic space downstairs, the architecture of the outside chapel was almost erased. In effect, the Catholic chapel, Our Lady of the Skies, was the rectangular base of this grand pyramid under the larger, more decorated Protestant chapel. Its ceiling was flat and unadorned. There were just small slivers of stained glass that received light only from the cement wells around the building, not from the sky directly. The front wall was a giant mosaic depiction of angels, and the simple lines of roof and walls added to the serenity without distraction. I thought it was absolutely stunning.

There were committee discussions before the women arrived about whether the academy needed female chaplains to minister to the female cadets, but the current chaplains assured the planners that they were trained and capable to handle counseling and tending to the cadets of both genders.[7] It would not have mattered to me as there are no female Catholic priests, but it could have been interesting to have one, the first one, there with us.

The chaplains were the good guys. They never wrote anyone up or corrected posture or decorum. As they looked out on the pews of cadets, they must have seen rivers of tears and many dejected faces invisible to the rest of the congregation who sat facing forward, but the chaplains continued to smile at us all. They reassured us that we could go on. They chanted, "Peace be with you," and "And also with you" jumped from my lips as an automatic response. I unwound in the pew just forgetting for the moment the similar but less peaceful automatic responses that the upperclassmen demanded outside those serene walls. The chaplains reminded us, in a very subtle and spiritual way, of bigger sufferings in the history of the world and in current times. My own concerns diminished while under that pointed, multi-spired roof. I sometimes smiled back.

I so loved the peace and quiet. I melted into the kneelers and seemed to absorb peace through the same kneecaps that throbbed with pain on the intramural fields or on the terrazzo after a slamming run to breakfast. My classmates, in fact, were out there right then doing exactly that to reach the goal of Mitchell Hall and some sustenance to start their day. I sat in the pew in serenity. I think I could have felt guilty, but the same religion notorious for instilling guilt let me off with my conscience unaffected on those mornings. I thought of nothing but calmness for the brief interlude while in the chapel.

9. My Darling Daughter

Mass was over way too quickly. After the final blessing, I slowly walked back down the aisle. As if by magical transformation at the threshold of the building, I resumed my stiff military posture, the shoulders back and down, the chin in, eyes facing forward and face stern. Once up the stairs, I stepped on the closest marble strip and broke into a jog. I greeted the same upperclassmen who just had shaken my hand and smiled for the sign of peace during Mass with a stiff "Good morning, sir." Off I ran.

Occasionally, I met a hard ass coming out of the dining hall as I ran in who felt he had to fill me in on some of the morning hazing that I missed. None of the fellow churchgoing upperclassmen bothered me at all. The glow from the windows and the prayers had not worn off enough yet for them to bark at anyone. It would later, though. I certainly couldn't count on a friendly face from the chapel to be the same kind face later in the day or on some other day.

One morning, there was a great snowstorm. I ran up the stairs from our room only to be greeted with pure whiteness through the glass enclosure of the stairwell's opening to the terrazzo level. I could not see anything through the heavy cotton snow blanket that covered the cadet area. I could not see the mountains as my guide, but I knew generally in which direction to go. I certainly didn't run or follow the marble strips on the perimeters. I couldn't see them. Struggling forward through the snow and against the wind, I couldn't even see my feet. I saw a dark shape, a gray ghost, in front of me. Thinking him an upperclassman, I called out, "Excuse me, sir. Which way to the chapel?"

"Hold on," I heard in return and took the proffered hand thinking of all the repercussions of being so familiar with an upperclassman. *I am going to get caught for fraternization with a cadet, and I have never even seen him.*

We found the chapel in the whiteness, and pulling back the hood of my black parka, I mumbled "Thank you, sir" before looking up.

"You are so welcome, my child. Thank you for coming out on a morning like this."

Not a typical upper-class response. Now out of the snow, I scrutinized the uniform to discover that it was in fact the hand of the chaplain that led me to Mass.

There were very few in church and Mass was no escape from the morning breakfast run. CLASSES HAD BEEN CANCELED: the only time in my four years at USAFA! I got the peace and support of daily Mass, but my roommates were still in bed when I returned. They escaped the run and the cold. It was the only day I could have slept in.

* * *

Spirit missions were also essential to our survival. Any prank-like show of enthusiasm and esprit de corps qualified. Spirit missions were partially about showing spirit to the entire wing, and under that principle, they were promoted and accepted. Spirit missions were also about keeping our own spirit alive and thriving in the tough USAFA environs. It is hard to believe now that after the rigorous days of running and training, after the yelling and corrections for all manners of minute infractions, we voluntarily left our warm beds to go out at night to run somewhere and chance another yelling.

"Get up. Get up, Kathy," Gail, Debbie, June, Peggy or Marianne whispered.

"No way," I grumbled, knowing that they would never leave me behind.

The only thing worse than being caught out on a spirit mission was for the rest of the women from the squadron to be caught out without you.

"I'm coming," I said and rolled out of the low metal-frame bed right on to the hard floor. "Just drag me along."

"Ha ha. Right."

We all began to don the clothes for the espionage of the night. This was no regular uniform combination. Doolies stationed in the halls in the mornings were assigned minute-calling duty. Starting at 15 minutes before required formations, they stood at attention looking at the clock mounted in the hall. At exactly ten seconds before each minute, they did an about-face and they announced the uniform for the formation and the menu for the upcoming meal. "Sir, there are 15 minutes until the noon meal formation. The uniform of the day is combination Alpha, wheel caps. The menu is…." Alpha indicated the more formal attire of our dress blues, something like a business suit, while Bravo through Delta were less formal. Then there were qualifiers for the fatigues, PE gear or the appropriate outerwear. But it was different in the middle of the night. The choice we made in the darkness for these spirit missions was an unlabeled, commando outfit accepted by lore and tradition to be the garb for nighttime missions. Some called it combination Zulu. Certainly no one announced anything in the hall for these outings.

For a stealthy look, we wore dark sweatpants and a dark top. Over uncombed and unruly hair that was just starting to grow out, we drew knitted black watch caps, low to the eyebrows and down over the backs of our necks. Black socks and tennis shoes for furtive walking finished this Zulu wear. We followed the point woman out of the room and on our way somewhere. Hopefully, the last in line would remember to not lock the door. Having to awaken the cadet in charge of quarters (CQ), the keeper of the keys, upon return would doom us all to exposure and demerits.

9. My Darling Daughter

More often than not, we did not join the men for spirit missions. The coordination was difficult. Missions were generally synchronized on a last-minute basis. If we had lived near our squadron, someone might have quietly knocked at our door. An exposed trip, all the way up to the women's area, though, added too much danger and lost sleep to a quick spirit maneuver. The men did their spirit thing, and we did ours, all in the name of the same squadron.

The objective of any spirit mission was a show of squadron spirit. We tried to endear ourselves to the ones who most closely monitored and controlled our lives. Our goal was to become a part of the unit, and that started mostly at the squadron level. We used skits and open jocularity to entertain the upperclassmen. We dressed as elves gluing cotton balls to the necklines of our gymnastics leotards to add a new dimension to the squadron's tradition of Operation Christmas, bringing gifts to underprivileged children in the area. During Parents' Weekend, Peggy, Gail, June, Marianne, Debbie and I rehearsed and then put on a skit in front of the parents

Parents' weekend, September 1976: Doolie women of 9th Squadron put on a skit for the parents. "Dress right, dress," Peggy yelled in her role as an upperclassman. Debbie in the back pulled on her bathrobe in mock ignorance of the command. Everyone laughed. From left: Debbie Wilcock, Gail Benjamin, Kathleen Utley, and Peggy Walker (author's personal collection).

Parents' weekend, September 1976. At lunch in Mitchell Hall with our families. No sitting in a brace or square eating that one day. Kathy Utley, Gail Benjamin (author's personal collection).

and the squadron. Peggy pretended to be a barking, angry upperclassman, while the rest of us played dopey Doolies. At the "Dress right dress" command, used in formation daily to align the rows, Debbie got "dressed" by pulling her bathrobe clumsily over her uniform. The parents got a good chuckle, and we showed the upperclassmen our willingness to make fun of ourselves even in front of our parents.

Outside of the squadron arena, there were other shows of spirit. Cheers from the jock tables by the rugby team or the intercollegiate basketball team erupted during lunches in Mitchell Hall. There were also class cheers with the entire Class of '77 rising during meals to shout, "Pride rides," their class motto. Cadets from our class might have sneaked out at night with 20 white bedsheets to form a giant '80 on the "Flat Iron" hillside above the cadet area. There was never a demonstration of women's unity. We did not want to identify ourselves by gender as a separate group.

The destination for the midnight forays depended on the purpose of the spirit mission. Our most common objective was to hang a bedsheet or two decorated with a catchy saying and picture in Mitchell Hall. Squadron spirit made visible by two white sheets pinned together painted with large letters, "GO NINERS" or "NINERS say: BEAT WYOMING!!!" With markers and paints and some of our own sheets, never slept on anyway, the artistic Niner ladies created different masterpieces. I contributed a

Parents' weekend, September 1976. Me as tour guide to show my family one of the challenges in Jacks Valley (author's personal collection).

little in the artistic department, but we all did our part with the risky foray to hang the finished banner.

No one ever signed the spirit signs. The idea was to place the sign and gain recognition for the whole Doolie group within our squadron. The perfect outcome would be when cemented at attention at the noon meal tables to have the cadet squadron commander notice the sign hanging high on the staff tower within Mitchell Hall.

"Did you hang that sign, Drewnowski?" he'd ask a male fellow Niner, with a voice warmer than the one he used for knowledge questions.

"Sir, I cannot say," Steve might answer with his eyes twinkling more than usual.

"You, Woodlands?"

"Sir, that #1, way to go, Niners sign? Sir, I do not know."

"You, Wilcock? Did you and your lady friends have anything to do with that?"

"Sir, would we be out of our rooms, after taps, against regulations?" she might respond, maybe with a little pride creeping into her voice.

The hassle-free banter continued with no one taking credit. If the show of spirit was funny enough to make the upper-class faces break from their normal serious casts when looking at us, or if it were original enough to get the moods of the upperclassmen elevated, table commandants might whisper an announcement from one table to another. Maybe, eventually, the group of Doolies from 9th Squadron might gain a release from the invisible spine-clenching posture with a smile and nod from those table commandants saying, "You Doolies, sit at ease."

Or it might take a bold Doolie asking in a booming voice, "Sir, in light of the awesome display of spirit on the staff tower, could the fourth classmen of the magnificent 9th Squadron sit at ease?"

A spirit mission sheet painted, then secretly sneaked to a prominent place for tradition and sometimes earning a little break in the training. From left: Debbie Wilcock, Marianne Owens, Peggy Walker, unknown, Kathleen Utley, Gail Benjamin (author's personal collection).

9. My Darling Daughter 131

No talking, when at ease, but the minor recognition and improved digestion would suffice for the positive motivation for that day. Sometimes the women earned a privilege for all the classmates in Nine. The men earned them for us too. Although exhausted from the night out and the adrenaline pumped during the mission, we exuded energy from the pride and satisfaction of having been part of the group and participating in the traditions.

We benefited from the laughter but no privileges once during football season when the men and women of Nine ran the ultimate spirit mission. We saw the final outcome in the squadron in the morning, and again I wished we lived in the squadron area to have easily enjoyed more of the mischief.

The cadets of West Point were in town for an interservice rivalry game, and the cadets of the 9th Squadron used the occasion to redesign the office of our air officer commanding (AOC) just down the hall. AOCs were the "real" Air Force representatives in each squadron. Officers of experience who supervised the entire squadron, they added the perspective of life after graduation to the training. They spent the day in the area with the cadets and went home to their families in the evening. In our case, however, this "real" Air Force officer was an Army officer, Major Ronald Sheffield. He was a veteran of Vietnam, completing three one-year tours, awarded many medals including the Silver Star, the Distinguished Flying Cross and two Bronze Stars. He accumulated over 2,200 hours of combat flying. He was qualified in six types of Army helicopters and four fixed wing aircraft. He served as my AOC of Cadet Squadron Nine and for other cadets at USAFA from 1975 to 1978.[8]

Major Sheffield often took cadets on motivational rides in a helicopter, heading north from the academy over the interstate and back around to the academy parade field for landing, flipping stomachs as he dropped the chopper that last 100 feet. He surely encouraged and inspired. This legend sent updates to our parents and kept us marching straight. He played commander and Dad at the same time or maybe alternately.

Years later when I asked him about his feelings on women entering the service academies, he did not hesitate. "I wanted everyone to be good." Good not in behavior, although I am sure he wanted that too. More like good in personal condition. He was opposed to making things difficult. Thinking his open-mindedness and open heart came from good experiences with women in his "real" Army experience, he explained that no, he had had little prior connection with women in the Army in airborne or in helicopters.

Major Sheffield, no matter how well loved, was the natural target for a spirit mission during the Army versus Air Force game week. Sometime after taps at 2300 and before dawn, the upper class and Doolie men of Nine moved his entire office right out onto the hill in the middle of the terrazzo. Everything, the desk and chair, the coat rack and bookshelves, books and memorabilia included. The cadets even placed the garbage can and small rug with precision to match their original position from the office. It was all there, right on the grass, in the forbidden zone for all cadets in the middle of the terrazzo. Of course, the job was not complete until a sign boasting of the perpetrators, not by name, but by squadron, "Viking Nine says: BEAT ARMY," was placed at the scene. They then filled his empty office with hay and placed a live mule, the illustrious Army mascot, there, in its new Air Force stall, for the entire night. June was a participant in the plan. "I was a farm girl and one of the 3rd classmen was a farm boy," she said. Somehow, they talked without getting in trouble and found a local farmer with a mule that he was willing to lend to them. The hardest part was that the mule had shoes on which made it slip on the floor. "We kept having to put sheets in front of him as he calmly went up the elevator and down the hallway. The mule was content eating and resting in the hay all night." June realized how fortunate they were that he didn't freak out and jump through the window. "The cool thing was, during spirit missions, relationships between the Doolies and upperclassmen were often relaxed. Yes, we were on the same team. And it was fun."[9]

Major Sheffield took part in the spirit of the day too. As he entered the office, we all gathered in the hall awaiting his shouts of disgust. We heard nothing. We looked at each other in confusion. A moment of murmuring inside proceeded a quiet releasing of the doorknob. Then Major Sheffield rode out, astride the mule, as if it were part of his daily routine.

"Good morning, cadets," he said smiling. "Beat Air Force."

10

Falcon Love

"I think probably kindness is my number one attribute in a human being. I'll put it before any of the things like courage or bravery or generosity or anything else."
—*Roald Dahl*

During our Doolie year, we learned and lived under the rules. There were books and books of rules and regulations to structure our uniforms, our behavior and our schedules. The first day in my room in Vandenburg Hall, I learned the rules covering the folding of underwear into a six-by-six square and placing the folded stack on the left-hand side of the top drawer of the dresser. The regulation book had rules for the correct placement of the bras to the right of the rolled stockings or pantyhose in the top drawer. The middle back of that same drawer was reserved for sanitary belts.[1] For men, the same position was reserved by regulation for their jock straps, also stamped with last name, first initial. When preparing for a SAMI, Saturday morning inspection, this same drawer should be open, according to the regulation, to an exact three inches with the one below open to six inches and the bottom drawer, balancing the look, open to a measured nine inches. Fourth-class cadets are allowed one off-duty pass (ODP) for permission to leave the academy grounds, per semester. All cadets, regardless of class, must sign in from weekday ODP privileges by 1915 hours and from weekend ODPs by 2300 hours. Name tags must be worn parallel to the ground centered over the left pocket. On the combination Alpha uniform jacket, which had no pocket, the name tag was to be worn level, on the left side, two inches above the top buttons, centered between the shoulder and the buttons. On and on, regulations defined the rules. All cadets were responsible for knowing the rules. There was "no excuse" for ignorance.

The fraternization rule became the topic of many discussions during our first year at the academy. According to the regulation, it was strictly forbidden for Doolies to have any personal relationship with any upperclassmen. The rule had been in existence for years. It was meant to be, and understood as, a rule to maintain discipline and a proper training environment. The no-fraternization rule was written to keep superior-subordinate relationships pure and uncomplicated. Without the added emotions of friendship, cadets would not confuse commands and favors. All subordinates were to be equal with no preferential treatment in the ranks.

Fraternization occurred, though, probably since the founding of the academy. Cadets were friends with other cadets of a lower class, even before the female hormones entered the mix. A fourth-class cadet who played exceptional football would be a trophy for any squadron-level intramural football team. That Doolie might not get the tough questions in ranks during the morning formation. The Firstie captain of the squadron football team might ask him, "What's your favorite NFL team?" The not-so-athletic Doolie to his left, however, might be the unlucky recipient of "Recite the entire inventory of attack aircraft, SMACK." Guys from the same hometown or younger brothers at the academy during the same four years might be invited to watch some TV before the regulations officially granted that electronic privilege. There were relationships, friendships between upperclassmen and underclass cadets, that compromised the clarity of orders and privileges before the women arrived. It happened. It was fraternization. But the rule and the relationships both took on new meaning with our arrival.

Initially, there was no real dilemma. The men were so busy adjusting to the women invading their domain that I think they, or at least their hormones, semi-forgot we were female. Many upperclassmen focused mostly on our role as intruders. Logistics were so mixed up with room arrangements, separate areas, open doors, and so forth that heads spun just trying to keep it all in line. The women represented troublesome Doolies and, at first, nothing else. "Now I need to wear a towel in the halls" Doolies. "May I touch you?" Doolies. "Are you dressed in there?" Doolies. "Sir, I am on my menses" Doolies. We were enough trouble with our new female Doolie requirements that there seemed to be no time or interest to jumble in any romance.

During BCT, the male cadets never looked at me in any fraternizing or compromising way. They looked to correct uniforms, to check knowledge or to demand responses. I had never received much attention from boys in high school. I preferred that passive ignoring to the demanding, often demeaning, corrective attention I received those first few weeks of Basic.

We were no sirens either. With our hair and confidence chopped to the roots, we did not exude great self-esteem. Makeup, civilian clothes, shoes with heels of any height went the way of our luggage on the first day.

Rapid showers left little time for leg or armpit shaving. Sun, wind, dust and stress lined our faces.

At first, I thought all the women to be kind of naive and innocent. I projected my own lack of knowledge and experience, when it came to matters of the heart, on all the other women like a kind of emotional uniform. Our blue skirts and boxy blue blouses hid my little girl figure and lack of knowledge about men. I just assumed that all the others were emotionally the same behind their blue clothes. I know I was wrong in that, but at the time, survival of the gauntlet of the terrazzo was so basic on my hierarchy of needs that I couldn't imagine for me, or any of the other women, any interest in the men except as tormentors to avoid. We all became less alike in that interest.

Soon, some female cadets were noticed. Even without the longer slit in her skirt or tuck in her fatigue pants, Gail got noticed. Sometimes, I'm sure, noticed for her seemingly rebellious challenges. Because she didn't cower, she sang out her own lyrics, and appeared certain in her purpose, male cadets looked twice at her. Debbie, another in our Basic Invaders group, attractive without mane or makeup, was noticed. One of those nice-looking faces that needed no coiffed frame or enhancements. Upperclassmen knew her name. Photographers pointed their lenses in her direction often. She bounced along seemingly almost always in good spirits and acted unaffected by the ravings around her.

For many of the rest, though, there were not even uniform combinations to boost the attraction. The uniforms designed by the head tailor could not compare with the peasant blouses or the miniskirts of the 1970s-era civilian girls. The choices in their closets made the female cadets all, well, uniform. Some wore the blue chambray skirt better, though. Some legs inherently looked nicer in the beige issue L'eggs brand stockings. Some could balance the midnight blue, felt mushroom-like beret with more aplomb than others. But at first, sweaty and dirt covered, wearing the fatigue pants and shirts from the assault course, we likely didn't draw any male cadet's attention. I thought that if we interested any at that time, it was because of our determination to be part of the wing, not any interest in being distinguished because we were of a different gender.

The men were not eunuchs. They were 17–23 years old, strong, intelligent, physically fit men who looked ultra-sharp in their uniforms. Their posture was perfect, standing straight and poised even when not at attention. They were clean cut. Most had standing haircut appointments at the barber shop once every two weeks. They drove nice cars, once they could. They definitely attracted the attention of many of the young ladies of the local area. They had girlfriends to meet them at the movies or restaurants downtown. They brought beautiful young ladies to the mandatory dances and functions. Doolie women attended in our uniforms and stood apart while the upperclassmen escorted their dates with the proper, white-gloved attention. Even our own classmates met local coeds from local colleges bused in to be the dates of the cadets for the evening. I remember coming out of a stall in the bathroom of Arnold Hall during one of these academy dances. The entire length of the vanity was crowded with young women in ball gowns of every pastel color. They stood shoulder to shoulder leaning in toward the sinks adjusting bra straps, applying lipstick and combing their feathered Farrah Fawcett–style long hair. Their chatter stopped as soon as my backward image appeared behind them. Two stepped apart so I could use the sink. After washing my hands, I retucked the white blouse of my mess dress uniform into the long black skirt beneath the silver cummerbund. I pulled the short bolero black jacket around my chest. No one said anything. I walked out feeling like an animal in the zoo.

In those years, I was envious of their freedom to dress any way they wanted. One ATO recounts that the female cadets were saying, "The girls in Colorado Springs hate us," and she said, "You bet your sweet life they hate you.... The biggest thing they want in their lives is a USAFA T-shirt with someone's name on it, and you ladies already have one, with your own name on it."[2] In similar bathroom encounters as an upperclassman, I had learned to appreciate being an insider, a cadet who knew and understood the system. Maybe they were envious of my connection to the male cadets or maybe of my confidence. I know I stood taller as an upperclassman and said hello to all of them when I adjusted my uniform in the mirror.

As a Doolie, though, it was hard not to compare outfits or hairstyles. I could see the results of all the primping and preparation that went into the beautiful looks they sported. Many cadets were a great catch even if they didn't get downtown often and could not have visitors in the cadet area. But squadron phone rooms at the end of the hall were filled in the evenings as cadets checked in with their sweeties.

A newspaper article that first year noted, "Some women cadets at the Air Force Academy feel they are losing their femininity in being treated like one of the men."[3] There it was, the femininity that the ATO warned Kathy Conley not to lose on her first day. I don't know who the reporter interviewed. We were harassed the day after the article appeared because of its mention that we might be missing bathtubs or kitchens. Luckily, no names of women interviewed were printed. Upperclassmen would have crushed them with persecution and "extra training." We did want to be treated like one of the men. That was the whole premise of the planning: acceptance through training that was as gender neutral as possible. But was it neutral, and how was it measured? By making the standards as close to equal as physiology and society would allow, the staff expected acceptance to come sooner. I wanted to be accepted as one of the cadets but as a female cadet. It wasn't about bathtubs or kitchens. It was some kind of femininity that the ATO described: something inside that can't be taken away—something about who I was. Issues of equality versus equity and inclusion versus integration within the existing system were still foggy concepts in my 1976-era 17-year-old mind.

The women overall wanted no differentiation. In her rebuttal to the original newspaper article, ATO Lieutenant Walter remarked that we "were incensed." She added, "They understand they are fourth class women and what that involves. They know they will have to go through doors last."[4] We knew, and we didn't expect upperclassmen to hold doors for us. Many days when all cadets streamed through doors of the academic building heading to class, I held the door and held it and held it. That was my job. That is what was expected of Doolies. I was a Doolie and expected to be treated like a Doolie, no more, no less because of my gender.

But we were women with varied levels of femininity. The real issue of fraternization likely began to germinate after the entire cadet wing returned and we all started the regular routine of daily classes. And hair dryers. And some spare moments to collect our thoughts and selves before facing the upperclassmen. We were a novelty; many of the upperclassmen were curious. They would stop us to vent some frustrations about our upsetting their routine and their academy, but looking at us and talking to us, they became aware that we were not so unlike their sisters. Maybe they stopped us because we reminded them of a classmate from high school. Maybe it was just nice for them to have a person to talk to with different sensibilities and sensitivities. At least, we were a change from the macho, hard-core, tough-guy attitudes that permeated cadet life before 1976. Surely not a welcome change for all, as the culture of those hard-core machismo attitudes was firmly embedded like the rocks of the mountains that surrounded USAFA. Those rocklike attitudes did not soften overnight when we arrived, but some male cadets started to appreciate the variety of sensitivities. Sometimes they may just have felt sympathy for us as we cried after a tough reaming out. Our emotion, mothballed during their time at the academy, let them see us in a different light. One classmate tells me that he would have eventually come to appreciate and admire the balance of female perspectives in his life through his wife and daughters, but not as soon as he did because of the presence of women at USAFA.

* * *

Once, the Niner ladies and I found a way to poke fun at the guys and their reactions to some female visitors. The cadets of Nine were great sports and most could laugh at themselves enough to realize that we too could laugh. The Playboy Bunny incident took them by surprise. Afterward, they saw us as fun loving yet determined, instead of withdrawn absorbing sponges waiting for their every command. We participated in the same game that they played but from a different angle. The result was a little chink in their armor and a night of hilarity and acceptance for us.

The Playboy Bunnies, the real Hefner, pointy-eared, long-legged bunnies, visited USAFA during the noon meal formation in the fall of our Doolie year. As usual, we lined up in formation ready to march. Unexpectedly, we heard the "Fall out" command when the "Ready march" should have started us, left foot first, on the way to lunch. During football season, it was not unheard of to have a quick pep rally before the formal marching. A parade of Firsties in their new, top-of-the-line sports cars would drive up the "Bring Me Men…" ramp and cruise around the terrazzo bedecked with banners proclaiming the greatness of the Air Force Falcon footballers. Normally, we all cheered and returned to formation to carry on with the march to lunch. This day, the Corvettes, Firebirds and Camaros drove up the ramp adorned not with signs but with live Playboy Bunnies. Perched like homecoming queens on the headrests of the seats or directly on the sacrosanct shiny backs of the cars, the women waved and smiled as the drivers beamed and cruised by the assembled cadets.

The bunnies were wearing their work uniforms, bustier-type, black, one-piece body suits with a plunging sweetheart neckline and high-cut legs over fishnet stockings. A large, puffy, white pompom tail was attached at the middle of the seat of the suit. Around their necks, they wore white detached collars with small, black bow ties in the middle. The outfit was sleeveless, but white french cuffs with black cufflinks circled their wrists like bracelets. Their headgear was the iconic black satin bunny ears rising above their long hair.

The men went wild. Catcalls and wild whistling filled the air. I wonder what the tourists gathered for the marching thought on that day. Maybe they thought those bunnies were the new female cadets. I was too small to see everything over the towering cadets in the front row. It all seemed absurd, but I happily stepped to the back of the crowd to enjoy the lack of training and attention.

Some of the female cadets were annoyed with all the commotion and that the men acted so enthusiastic over the bunnies. I was only partially offended in the women's libber, equal rights–type way. I did appreciate the moments of anonymity when no upperclassmen were looking at me for corrections, but it did seem ultra-ridiculous that the men were making such a commotion about black headbands with ovate protrusions. It was like they were World War II aces flying in from some dangerous missions and the personification of the nose art pinup girls from their airplanes were there in skimpy outfits to greet them. My own mind was churning trying to come to some balanced understanding about me, of women, in this men's world. I was still young, but my female classmates and I were women. Were there different levels of womanhood, like hormones? Did the clothes make the woman? Did sexy outfits make them more feminine or fatigues make us less so? The bunnies sat on the staff tower during lunch and caused a hullabaloo again when introduced to the wing as special visitors.

Before the evening meal, Peggy, Gail, Debbie, June, Marianne and I discussed the issue.

"It's funny to see these macho men reduced to Jell-O over a few black ears," June commented. June was older, having been at Colorado State University for three years. She was not as awestruck by cadets as I was.

"They couldn't talk about anything else at lunch. We'll probably hear it again at dinner tonight," Gail replied.

"We should show them and dress like bunnies ourselves," June suggested.

Peggy seconded the idea. Peggy and fun got along well.

"Oh sure, let's just search through this wide array of clothing choices and pull out a Playboy Bunny outfit," said skeptical Marianne.

But that didn't deter us.

"It's so ridiculous, so let's be ridiculous and show them," I added, now caught up in the fervor of the idea.

And so we invented another new uniform combination never used before, and I'm sure never ever duplicated: combination bunny.

We started with our athletic shorts, a nylon navy blue short with about a three-inch inseam. We put our ugly light beige, not at all fishnet, nylons underneath. We owned nothing showing any cleavage. Our regulation two-inch-spaced hangers held choices ranging from long-sleeved, olive drab, canvas fatigue shirts to light blue boxy blouses with a small blue tie tab snapped at the neck.

"Let's just wear our swimsuits and get on with this," said Peggy as she pulled issue blue grandma suit up and on.

We added our no-heeled, calf-high black leather combat boots. If the uniform committee before we arrived had gone with one of the original options, we would have had white go-go boots in our closet. It would have been perfect for that one occasion but not worth it for what would have been a harder day-to-day acceptance battle. But we did have issue black combat boots. Boots well broken in by over three months of wear. At least they were shiny with the spit and elbow grease of many hours of polishing. So far, the outfit we created was only completely bizarre and not at all bunnyish. It didn't matter by then. We were playing and mocking the machos.

Someone's, probably June's, creative juices supplied the idea to use our parkas. No ski bunny, tight-fitting, butt-hugging parkas, these were square, black wool garments, not a dart in them. They would have zipped up over a TV box and looked as becoming as they looked on any of us. The hood, also stiff, black, unadorned wool butterflied open like a pork chop in the back to lie flat on the upper back when it wasn't terribly cold or zipped up over the head for those wind-chilled Colorado days. The parkas were longer than the shorts, which actually worked, making it appear more like a mini dress than a parka.

"But the ears. We must have the ears. If we go out there and there is any doubt about who we are trying to be, they'll eat us alive with questions and we'll never get to training on time," I said.

"Voilà! Ears," exclaimed Debbie as she zipped the hood down only halfway and the two unzipped top halves of the hood points stood up. We pushed our felt berets, which usually bobbed alone on our heads, under the points. The berets worked perfectly to support the wobbly flaps.

"This will do. We are going to be late," someone said, nervous about breaking regulations and getting in trouble.

But we were not bunnies yet. Something, a rolled-up sock for one, a wad of cotton balls or a crunched piece of paper for others, all held with a giant safety pin from our green issue laundry bag, finished the back of each costume with a white tail.

"Perfect," we yelled. We jumped and laughed and hugged each other.

"I am not hopping to the squadron. I am all for a good laugh, but my legs won't take that," said a smiling Peggy.

It would have been early in the evening when the peaks of the Rampart Range immediately to our west would blacken against the multicolored sky. The mountains were a jagged curtain that blocked the sunlight long before the sun really set on the horizon. We blended in with the darkness in our "combination bunny." That spirited night, in semi-obscurity, we headed out across the terrazzo and down to 9th Squadron as a line of black crazies yelling, "Good evening, sir. Niner bunnies."

Our posture crimped forward after each outburst to control the laughter. In the twilight, we must have looked like some kind of bowing monks of the deity Big Ears.

The upperclassmen laughed, too. They first did a classic neck-springing double take. "Good evening, sir" hardly registered with them. Many didn't even turn to look. The "bunnies" part, with the white tails bobbing off along the marble strips, got their attention, though. A few stopped us with seemingly feigned anger to ask what had become of the precious USAFA uniforms. "What's happened to this place? I told you it would never be the same once the women were let in." But their own classmates smiled and cheered. "Hey, it is better than having the Doolie guys dress up like bunnies. Very creative, Niners."

They laughed and not at us but with us. Not a one stayed angry, at least not enough to punish us. It was spirit after all.

* * *

The confidence of starting to belong must have made us more attractive in the nonphysical sense. When we actually thought, acted and spoke on our own, we were a group of well-educated capable young women. The men started to notice us more and we noticed them. For so many reasons and with a ratio of 40-1 men to women, the rules about fraternization were soon overshadowed by hormones. But they were not totally forgotten. There were still men who were adamant about getting rid of as many women as possible. Vigilant upperclassmen were on the lookout for violations. Finding a female cadet fraternizing would be a serious black mark for that woman and cause for some major punishment for both participants. Most of us ended up with dates with upperclassmen but not all. "Some hate 'em, most date 'em" was a murmured motto for upperclassmen.

Captain Galloway, the female officer who worked very closely with the planning committee in preparation for our arrival, admitted that one area of concern for the committee was fraternization. Galloway explained in her interview that "the integration of the cadet wing is first and foremost a social phenomenon." She mentioned that the physical, emotional and mental ability of the women to meet the demands of the program were "never seriously in doubt in the minds of the planning staff." She did think that the issue of how the women would adjust to the social pressures was a concern.[5]

Karen Wilhelm describes an incident with the female cadets in the Drum and Bugle Corps: "For the first away football game, the powers that be were having kittens thinking about six women traveling with over 100 men, staying in motel rooms, and good heavens, 'What might happen?!!'" Those powers decided to send one of the ATOs on the trip to chaperone and keep the women cadets out of trouble. The ATO brought the women together for a meeting and suggested, as Karen tells, "that if we felt the need, she could get us an emergency appointment at the cadet clinic to get set up for birth control!" Karen was fairly certain the idea wasn't the ATO's. For Karen, it was another of those "'what the hell?' moments." Walking down the hall away from the ATO after her meeting, one of the women exclaimed, "What do they think we are, a bunch of sluts?" Karen remembers, "Somehow, we made it through the trip without any Drum and Bugle babies being made, and the ATO chaperone detail was discontinued pretty quickly."[6]

Little hidden relationships started with some of the women. Gail noticed a tall, handsome, Three Degree on the terrazzo one day. No need to ask his name. It was right there on his jacket. His "address"? Right there, too, with a patch for the squadron that housed him. Did she then look for him among the multitude? Yes, without "gazing around." Did she wander casually through his squadron hoping to see him? No, that would have been too much. "What are you doing here, Benjamin?" "Are you looking for more training, Doolie?" A whole new set of upperclassmen to answer to? No, probably not. Did they meet on the athletic fields? Did she purposefully violate some regulation hoping he would "call her out" and then ask her out? However it happened, many of the women, myself included, ended up with upper-class love interests who gave us support and escapes, while adding extreme drama to the days.

My first cadet boyfriend was an upper-class co-member of the pistol team. The relaxed atmosphere in the field house at the shooting range gave everyone a chance to communicate beyond the five responses. We couldn't always talk over the blasting of the pistols and rifles, a weird environment for budding romance. Funny, those gunshots all around were more peaceful than walking across the cadet area on the terrazzo. But we did talk some. "Nice grouping, Utley," he might say for a ring of bullet holes near the bull's-eye. A quick "How's your day, Utley?" at an afternoon practice from the same upperclassmen. He had walked past while I got chewed out on the terrazzo for not saluting the officer of the day, but then a mere compassionate-sounding hello could melt my heart in those days. Anything said beyond "get your chin in, your shoulders back and down" was a joyful noise. It didn't take much wooing to get things started when I was so vulnerably hungry for kindness.

Our first date was in September or October 1976, five months into our journey through USAFA. The repellent effect of our intruding on the men's space was wearing off some. The team member was a Three Degree, a member of the Class of '79, a young man alone far from home too. He just spoke to me, and I was interested. I couldn't wait to get to practice just to have his smile or "hi" greet me. We finally felt brave enough to thwart the regulations and get away for a "date." We could not go anywhere, as I had no privileges to leave the base. Even if I had, he had no car. Car privileges were awarded to only the juniors, at the very end of their Two Degree year. Chris Young, the officer who had picked me up at the airport my first night in Colorado, had a VW bug that occasionally got me wheels on loan. Sometimes if the football team won, the whole wing received the escape of an off-duty pass. That translated into freedom from after the game, 1500 until sign-in at 2300. Another female classmate and a team member were going out, fraternizing too. We'd double date if they had transportation. We managed to escape-date a couple of times.

10. Falcon Love

You could not meet a romantic interest in the halls and strike up a conversation, nor could you chance getting caught together anywhere in the cadet area. Men and women in the same room, no matter what class, were required to keep the door open. No phones in our rooms for any cadet meant no easy communications. To meet, you would have to steal away. We had nowhere to go and especially nowhere we could go and not be seen.

There was also no way to disguise ourselves. Surely, the female cadets had more ways than the male Doolies with their shorn heads, but no Doolies were allowed to wear civilian clothes. To go downtown in uniform would announce to everyone the fraternization with rank right on the shoulder boards. The name tag with the names of the perpetrators, for easy accusation. No, we couldn't easily go downtown to get together. Plus going incognito, in mufti, magnified the punishment, breaking another regulation while fraternizing. During the course of a felony or something like that. But we risked it.

We schemed and planned to rendezvous. I cannot fathom how I had the time. We saw each other every day at the range but could only act like teammates. I wonder if we fooled anyone. The teammates looked the other way. I kept shooting well and worrying less, so maybe for the good of the team, no one said anything. We exchanged a few notes at the pistol range, but despite the thrill of the clandestine meetings, the spark wasn't there.

I dated a cadet in my own squadron next, and new spy-like maneuvers added to the thrill but also to the strain to not be discovered. To stay connected, we exchanged notes and letters directly from cadet post office box to cadet post office box. That was probably another violation of federal postal codes or something like that. The infractions multiplied. We carefully watched for the box neighbors who might realize that you were opening a box that was normally visited by a taller, more senior, male cadet.

I'd open the small brass door of his mailbox with his secret combination and deposit my note. Then I would casually, discreetly wander over to my mailbox and check inside for the folded 8½ × 11 piece of lined notebook paper, no envelope, no stamp, that would add a "can't wait to see you" to my day. With these strange communiqués without envelopes appearing between sorting times, the postal clerks must have known what was going on. I wonder what they thought of all of it. I wonder if they told anyone in our chain of command. Maybe the commanders knew and ignored the whole occurrence. In the grand scheme of integrating women into the cadet wing, passing notes to upperclassmen might not have even registered. But cadet regulations are not Air Force regulations and definitely not federal regulations. The mail clerk probably told no one and really didn't care. It might have been minor to him, but for me as recipient, the relief from the discipline and the energy from that fuel for making it through the day, was major. I tucked those letters between pages of my chemistry book or military studies text and marched all the way back to my room and shut the door before even reading one word. With those notes and the relationships with the male cadets as boyfriends, I softened. I found my smile again. I felt like there was a way here. Cadets were okay. Just dating one made the rest seem human. Well, not all the rest but many.

After Thanksgiving when we got phone privileges, my post office box notes to my cadet boyfriend sometimes arranged for evening calls. *I'll call you at 2030 tonight, Kathy.* Then at 8:30 p.m., part of cadets' unscheduled time when we were supposed to be studying, I walked down to the squadron, chin in, shoulders back and down to the phone room and slipped into one of the booths. Forgoing the quarter-round cantilevered seat in the corner opposite the door, I stood, back toward the door, for added privacy. Finding the right voice was critical. Not too loud lest the neighbor callers heard. Not too softly so I wouldn't have to repeat. For those like me, dating a cadet in your same squadron, a weird spy movie scenario resulted. I called the phone booth next to me and let it ring until a passing cadet answered. Then in my disguised sweet civilian girl voice, with mouth and eyes hidden, I asked for "Cadet First-Class Someone, please." I sat still and moved my mouth pretending to talk to someone as I waited for my guy to make his way to the adjacent booth. I'd notice him sit down on the other side of the shared booth partition talking as if to someone else. The conversations were typical sweetheart calls with compliments exchanged and news of the day. The clandestine arrangement, the shared phone booth wall, the romance and the fear of discovery was certainly a new USAFA occurrence.

*   *   *

Women were first admitted to Yale College in 1969. Like us, those first women stood out. One female Yale '71 grad, a lawyer in Washington, D.C., said such attention had its advantages and disadvantages: "We were on the spot all the time. But look at it the other way, we were invited to every party."[7] Advantages and disadvantages: the same held for us. We had advantages of getting cars to use. Male classmates planned weekends around going to cadet hockey games and movies on campus at Arnold Hall. After winning football games, with passes in their pockets, they had no means of escape but saw their female classmates leaving happily with keys to upperclassmen's cars. It was not forbidden to borrow a car from upperclassmen. Upperclassmen just didn't loan out their new Corvettes, Firebirds and RX-7s to anyone who asked, especially not to Doolies. But the Doolie women got them. Some of our classmates harbored resentment for a reprieve they would never receive because of their gender. I never knew it until later. I would have taken one along if they had asked. Why didn't our own classmates ask us out? Did they fall for all the rhetoric of what misfits we were? Were they still haunted by the Janes they had left behind hoping to reattach themselves to them and their previous lives at the first break at Thanksgiving? Did they just not see us as feminine? Did the women feel the same class relationships lacked the excitement of breaking the regulations to keep them interesting? Did we ignore them for the adventure of a clandestine relationship with a cadet with a car?

After our Doolie year, fraternization was no longer an issue. The rule did not apply to Three Degree relationships with Two Degrees or Firsties, although one could technically outrank and command the other. The regulation applied to the situation of one having extreme power over another—superior-subordinate. Certainly, the case with a Doolie and any other class.

By the time we were Three Degrees, we knew how rank worked and knew that, although they outranked us, we were not under the strict control of the more senior cadets. We knew enough of how things were supposed to play out to know that no one could order us to do much. We had the chance then to become friends with our classmates and the members of the Classes of '78 and '79 as much as they were interested.

But as Doolies, because of our romantic relationships, some female cadets, myself included, got breaks that we didn't deserve from the training. Some of our classmates noticed and resented it. They were already tired of all the press and the headlines suggesting that the Class of '80 was made up of women with a smattering of men added in to fill up the spaces in the columns of marchers and the photos. My upper-class boyfriend was best friends with another cadet known to be an extreme tyrant and no advocate for women at USAFA. I was afraid of him and afraid of what he would do if he found out I hadn't shined my shoes, let alone that I was dating an upperclassman. But in one of the secret missives from the illicit post office box exchanges, my upperclassman said not to worry about the tough guy; he would not be a problem. I got immunity by osmosis. I know any classmate having received the tirades of that strict cadet would resent that I walked by unnoticed and unthreatened if they knew why he ignored me. Did the added harassment by unknowns just for being female offset the leniency and added privileges of these relationships? I know I spent more hours being yelled at for being an intruder than any hours treated with special privileges. I know I got upperclassmen's cars to use to get away but that the obstacles on the confidence course gave me no confidence because they were not built for my petite body. I couldn't reach the beams to swing up and balance with surety on top. I know that I feel equal to my classmates now. I know that I lent my car to male and female cadets when I made it to that upper-class privilege. But I only know my mind. I cannot by proclamation change the minds of others who saw inequities and favoritism and cried foul. To create a 3D picture, there must be two different angles.

We dated some cadets but also became "one of the guys" with many male cadets of all classes. I had great times with the guys as friends and loved many non-romantically. They were my Falcon brothers, and I trusted them. We shared the bond of training, of hardship, of inside jokes and "war stories" that outsiders would not understand. We had secret vocabulary for both straightforward and difficult situations. We sat and talked for hours about their girls back home, problems with classes and plans for our careers in the Air Force. We went out together to movies or for ice cream, each paying our own way or pooling money from our almost-bare wallets. We skied together on weekends. So many of the men held us up. They whispered or yelled support. They encouraged us, stood by while our pictures were taken.

Marianne remembers walking back from the library late one night as a Doolie. It was almost taps. It was pitch black and she heard, "Hey, cadet! Female cadet." She knew he was talking to her as almost no one else was on the terrazzo. "Right face!" he said, and she thought, "Right face? If I turn right, I am going to look out over the forest not at him?" But she dutifully turned right as he walked up behind her. She recalls, "There was this moon rising and it was a huge harvest moon, just beautiful." And he said, "Didn't want you to miss it. Carry on."[8]

There were some amazing men in our class and at the academy to help us along the way. Thanks to all of them. The men of my squadron particularly were wonderful and supportive and concerned. They were willing to define cadet to include me and what I brought to the relationships. We were not fraternizing. We were surviving. Together.

Later in my cadet career, the issue of my Doolie fraternization resurfaced, long after the statute of limitations on my fraternization violations had expired. But during my Doolie year, I dated upper-class cadets and was never caught or punished. I certainly gained some freedoms from the relationships to get away, to leave the harshness behind and to see upperclassmen as men and not just harsh trainers. I broke a rule, which was chief to my understanding that try as I might, I could not be a perfect cadet. There were major rules, such as the honor code and the forbidden drug and alcohol use in the dorms, I would honor. To me, these seemed central to the academy and what it stood for. But fraternization did not make me ineligible to be a cadet. It did not make me unworthy of officership. I was vulnerable and broke the rules and, with massive disappointment, I would have taken the punishment if caught. But I found a human side to the upperclassmen of the academy, which was even more significant.

11

Black Panther

"You may have to fight a battle more than once to win it."
—Margaret Thatcher

Through the first semester, even from the isolated women's area, we female cadets incorporated ourselves into as much cadet life as our separate housing would allow. We learned the names of our squadron cadre and got to know our squadron classmates. We did all that was required and more to prove ourselves worthy of being part of the squadrons and the groups. Some, however, required no proof. Others would never be convinced. But we settled in.

We still had ATOs watching over us, but we tried to stay close to the male classmates in our squadrons, hoping to still some of the claims that we were getting special treatment. We grew close to achieving some

Crazy Doolie days. Hanging together after 2015 hours when we could be out of daily uniforms but in uniform PJs. From left: Marianne Owens, Susie Park, Kathleen Utley, Gail Benjamin, Peggy Walker (author's personal collection).

acceptance when, at the end of the semester in December, a decision came down the chain of command to go ahead and integrate the women fully into the entire cadet wing. To rectify what was called the "Terrazzo Gap," the squadrons that had women versus the "have-nots," now squadrons, numbered 20–40, mostly housed across the terrazzo in a separate dorm, would get Doolie women in their ranks.

Alleluia!! It was about time. If women were to become part of the cadet wing, then from the first day we wanted to be in every squadron, the squadrons in Vandenburg Hall on the north side of the terrazzo and the squadrons in Sijan Hall to the south. From the beginning, there should have been women on all squadron intramural teams. Female cadets should have eaten at every Mitchell Hall table, breakfast, lunch and dinner. We should have run night spirit missions from Squadron Twenty-Five Redeyes, Thirty-Three Cellar Rats and all the others up to Ali Baba Forty. In 1954, in a letter to retired general W.W. Welsh, the very first academy superintendent wrote, "The Air Force Academy Act prescribes that our institution shall be for the training of 'selected young men.' If you can get the law changed, I shall be delighted to take in young ladies as well."[1] Twenty-two years later, the law was changed. We had been admitted but not totally "taken in." Finally, someone was initiating an attempt to right the wrong, to even things up, to make our admission complete.

Superintendent Clark, the superintendent who oversaw most of the planning for our integration but left before we actually arrived, admitted, "One of the most significant mistakes was putting the women in separate barracks. It isolated them from their squadron mates. So many things go on in a barracks that you need to participate in to become bonded with the rest of the squadron, so you don't feel like outsiders."[2] From the vantage point of 50 years away, he could see that the isolation would not help acceptance.

What we didn't know was that the time we spent isolated in the women's area bonded us to one another as no other class's female cadets have bonded. Even though we hated the separation, because we couldn't connect with all of our male classmates as academy training demanded, we relied on each other, across squadron lines and between rooms all up and down the hall.

The protection was a factor too. At 17, I hadn't realized I might need protection. Should we fault the academy for wanting some semblance of security in their plan? We know now that physical separation is not the answer to the problem. Walls protect, but with them in place, there is no inclusion.

But in 1976, I think all the women agreed with the concept of complete integration. We realized the necessity of moving female cadets into the 20 still-all-male squadrons. We all knew it was for the good of the integration, for the good of the academy, for the good of the women. Major General Susan Desjardins, a classmate and former USAFA commandant, relates that she was excited and nervous for the integration to occur. "I think the administration saw the barrier of separate living and made the decision to integrate us to help with squadron unity," she said. She also recalled that the ladies would "wake up and do details [chores] for the women's area and then go to their respective squadrons to do their daily duties like minutes and trash detail. In many ways this arrangement made life for the ladies harder, so most were excited to finally move into their squadrons."[3]

She knew. I knew. We all knew. All semester, I wrote letters home complaining that that full acceptance would require full integration. I had tried to convince family and friends that living within the squadron, with our classmates and all the upperclassmen of that squadron, was essential to our being accepted. I had not convinced myself enough, though, to be an excited volunteer to move to the other side. I certainly didn't yearn to be a pioneer integrator again. I was settled, semi-happy and established. I did not want to be among the "going." Even though we knew it was for the best, none of us was enthusiastic to be the ones to move.

It was going to be hell all over again. I knew some of the male Doolies on the other side of the terrazzo from our BCT, Invaders squadron. I thought they would be allies in the foreign land of Sijan Hall. But in a way, they had joined the boys' club. They spent half a year living without women in their midst. Marianne told of walking into that men-only zone during the first half of the year to visit the tailor shop in Sijan Hall: "Like the Ho Chi Minh Trail … horrible names yelled out at us and we'd be hauled out and yelled at made to do push-ups, just trying to walk to the tailor shop."[4]

We would have to win those classmates in the male-only squadrons over along with all the upperclassmen. Some knew us from academic classes. It was entirely another thing, however, to have women in the living areas. They were losing the luxury of "towels only" to and from the showers. Losing the athletic field dominance with no women on their intramural teams. Losing the male bonding with no women at their tables or in their squadron formations. Things were going to change.

The air officer commanding of each squadron made the terrible assessment about those to go and those to stay. Susan Desjardins reminds me that some volunteered and some were volun-told. She did not move but recalled, "I know I looked at those who moved as the strong ones, academically, physically, those women who were meeting the challenges and would be successful with a pretty dramatic change." A different female classmate who was moved complained about the move: "Right when we were getting to know each other, those ties were broken." Her AOC announced the move "cold-heartedly," she remembered. "It was sort of like I knew the guy didn't like me, and I thought he was trying to dump me off on some other AOC because he was trying to get rid of me."

Our wonderful AOC, Major Sheffield, had to make his own difficult choice. Of the six women of 9th Squadron Doolies (June, Marianne, Debbie, Peggy, Gail and I), three would stay. Three would go.

We all sat in one room awaiting his decision. It was one week before final exams. We were nervous. I felt closer than ever to these women. We had bonded as much as we could while so busy trying to be a part of the entire wing. Now I felt the tug of those bonds breaking as some of us were going to move too far to share stories or spirit missions. We sat close together. We held hands like the finalists of some beauty pageant.

Major Sheffield spoke to us about who would move to 29th Squadron. Peggy, Gail and I were the chosen to become Black Panthers of Squadron Twenty-Nine. When the silence after his words struck some with relief, others with anger or fear, I did wonder why I was chosen. Did this fatherly major think I had the guts to start over and win over a new group? Did he balance two teams of players by fortitude, academics and military knowledge and give one team to each squadron? Did he know I was dating an upperclassman from my own and his squadron and elect to use the expanse of the terrazzo to keep us apart and save me or him from getting caught? Many years later, he admitted that he chose to send those he thought would handle it best. How could he have seen that in me? One of the 29th Squadron Doolies told Peggy they had a squadron meeting before our arrival. The entire new squadron was told that they were "getting one on Superintendent's list [awarded for military and academic excellence], one on Commandant's list [awarded for military excellence] and one troublemaker." Since Peggy wasn't on either commendation list, she knew she was the "troublemaker." I'm sure she thought, *Bring it on*. When we arrived in Twenty-Nine, before we had even interacted with anyone, an upperclassman came into our room and announced to Peggy that he had taken it as his personal task to reform her. Bring it on!

Our classmate Susan Desjardins reflected on the policy: "I thought it was the right thing to do. Something had to change. The women cadets were not making the strides we needed to be making in being accepted, respected and recognized for who we were and what we could contribute." She appreciated that the officers in leadership positions saw the ramifications of the division and adapted the original plan to fully integrate all squadrons. She "applauded the leadership for listening to the women and the men. We started beating this drum pretty early on and the change happened!"[5]

Life with the Black Panthers in our second Doolie squadron started after our Christmas break. When finals were over, studied for with troubling and distracting thoughts about the transition, I went home and enjoyed the warmth of Florida and of my family. I was sick about going back. I knew that when I returned, I would be standing at the base of the "Bring Me Men…" ramp again. I didn't want to start over. I wanted the security of knowing the cadets in 9th Squadron, knowing which upperclassmen would explode over unlearned knowledge and which were the kind of in-ranks inspectors who would scrutinize every inch of my uniform looking for an infraction. I wanted the comfort of the phone booth routine for calling my cadet boyfriend and knowing that although I could not look at him while we talked on the phone, his shoulder and mine rested on the same partition between the phone booths. I needed the predictable comfort of the layout of the squadron, the distance to classes, to the athletic fields and to formation.

The Black Panthers of Twenty-Nine knew nothing about us. Twenty-Niners did not know we once teasingly dressed up as Playboy Bunnies. They didn't see our skit for Parents' Weekend when we happily made fun of ourselves to make a place for ourselves in the squadron. They didn't know we would try to be in the squadron as often as we could, that we wanted to belong. But they also had never seen our mistakes, although our records had clearly breeched the gap. There were so few of us women overall that I had to believe that beyond which uniform pins for achievement we wore, they all knew more things about each of us.

Still, we knew the nomenclature and we survived BCT. Our rote memorization of the pages of *Contrails* would go with us well etched into our gray matter. I had an upper-class boyfriend whom I would continue to keep secret, but I would no longer see his face daily from the other side of the terrazzo. Maybe that was a good thing to keep us from being discovered. I had Peggy to make me smile. I had Gail with her confident ways and snappy answers to keep things in perspective. I was comfortable in my uniform, even the now well-softened and highly shined leather of my combat boots. We had won a place in 9th Squadron. Could we do it again in Twenty-Nine?

In January 1977, fellow Doolies passed our belongings across the terrazzo in a process that reminded me of an old Cold War prisoner exchange. Like a scene from a black-and-white movie, I played one of the prisoners marching across a bridge dividing the two enemy territories with each side meeting at the center. The Doolies from squadrons 1–20 pushed metal laundry carts holding our hanging uniforms and toted green laundry bags filled with the contents of our three dresser drawers and small overhead compartments. In a repeat of the inprocessing day one, we also carried our rifles, our boots, our nagging fear. We met the Doolies of squadrons 21–40 halfway across the terrazzo. These new squadron mates accepted our things and us in a somewhat dejected manner. Nothing was passed back in the opposite direction except maybe some glares and snickers as both sides of the terrazzo now included women.

Even though we now were part of every squadron in the cadet wing, we were not fully integrated. The new women's area on the Sijan Hall side was a separate area too. Again, the women would supposedly live separately but with equality. We would have staged our own 1960s-style lunch counter sit-in, but we were in a totalitarian state.

Peggy, Gail and I found our room one floor down from the terrazzo. We quietly opened the door to 2D27. We slid our blue-and-white plastic name plates in alphabetical order into the metal racks by the door. Benjamin, G.F., Utley, K.M. and Walker, P.C. in the last room on the end of the women's enclave became the three women of the 29th Squadron Black Panthers.

When scoping out the quickest way to the squadron, we squealed, much to the chagrin of the upperclassmen walking by, to find that providentially our room actually abutted our squadron. *Our* squadron area. Our wall was, in fact, a 29th Squadron wall with a couple of Black Panther Two Degrees living on its other face. Sure, a set of privacy doors separated us from the rooms officially assigned to the 29th Squadron's Black Panthers, but those doors didn't lock and remained open during daytime hours. They were no physical deterrent to our belonging.

We jumped right into squadron life. We learned all the new upperclassmen's names. We fought back more angst and clenched our fists while at attention as they challenged us to "get with the program" and "get up to speed." But we waited for evening call to quarters' knowledge training in our own rooms now, getting some homework done or shoes shined. Our peers came to our rooms to visit, to plan mealtime cheers or spirit missions with us, rarely stopped by an ATO as their desk was at the other end of the hall. Being adjacent to our squadron gave us privileges of belonging that we hadn't even known existed. The acclimation was easier because the wing had somewhat accepted that all the women were not going to quit this first year, that female cadets would be part of the future of USAFA. But it was harder because this squadron of Black Panthers was already filled out with the male Doolies. We had to methodically scrape out a place for ourselves.

We learned a new skit for a squadron meeting. We practiced in our room quickly lowering the volume on our cassette player or hiding in the closet when upperclassmen knocked to find out "what in the heck are you ladies doing in there?" Finally at the weekly squadron meeting one Thursday night in January, after the evening meal, we stood in front of the men of Twenty-Nine. After all the formal announcements about keeping the squadron running smoothly for another week, a fellow Doolie announced Gail, Peggy and me: "Gentlemen, I give you the Inverted Oreo." We stepped in from the wings, Gail in the middle, the chocolate cookie center between the white frosting of Peggy and me. I saw cadets in the room squirm and some upperclassmen's eyes roll sideways to exchange disgusted looks with male cadets seated to their sides, as the music from Barbra Streisand on our *A Star Is Born* tape boomed from the speakers. We started the routine, well rehearsed, to the "Black, black widow is sitting in the middle of the web it's a fly she seeks...." Gail crept up between Peggy and me flailing her arms as if she had eight. Peggy and I danced mirror image backup moves singing the "ooh, oohs" from the chorus. Slowly, some faces in the crowd broke from stiffness to slight smiles. Whether entertained or aghast, we didn't know. I'm not sure we cared. It was academy tradition to entertain the upperclassmen at the squadron meetings. We were Doolies, and we were doing a Doolie skit. We were women and made the skit our own. We would be Black Panthers in the demanding aspects, military and academic. We would be Twenty-Niners in the fun too.

* * *

Moving to the other side of the terrazzo, we quashed some of the speculation that women were receiving different treatment, positive or negative, but we had not yet escaped the ATOs. ATOs were still involved in our everyday activities as stand-in upperclassmen along with the actual upperclassmen. Half the ATOs followed the women across the terrazzo, but now we shunned them. We wanted our independence. They too represented the separate treatment that we didn't want.

Still, at taps each night, the female ATOs visited each female cadet room, counted heads and beds to assure all were accounted for. It was a very perfunctory dormitory inspection routine and one that we knew well. Once, at just minutes after taps, an ATO popped open the door and found us still not completely settled, lights not out. She signed a punishment form-10 giving us each demerits for being up after taps. By less than five minutes. But according to the regulations....

Now in Twenty-Nine, though, the upperclassmen assigned dormitory inspector (DI) duty from our squadron also visited our room just as they did for all the fourth-class men. Cadets, not surrogates, they added spirit and relief to their visit. Dressed as Hare Krishnas wrapped in red parachute material, the DIs chanted good nights and held aerodynamics texts as holy books. They came in wearing the attire of comic fighter pilots, aviator glasses on and cigars hanging from their lips and ranted about bogies in the skies, then headed to the other Doolies of Twenty-Nine to continue their emoting and counting. They laughed and we laughed. Many nights, upperclassmen DIs appeared as other characters or themes as they checked off the head count at taps. In accepted tradition versus strict regulation following, the minutes after taps didn't figure in the process.

Our classmate who lived across the hall from us in the women's area of Sijan Hall traveled up two flights and over one quadrangle to join the men in her squadron. She felt for us, though, as the ATOs seemed to resent our quasi-incorporation into squadron life without their help. Perhaps they resented the intrusion into their domain by the men of Twenty-Nine. It must have been confusing for them. The upperclassmen practiced so many unwritten rules. The ATOs had only their shortened mock cadet experience and the regulation book as guidance. Some ATOs seemed always ready to hand out demerits for things that were noted as infractions in the regulations but traditionally left unnoted.

To thwart one ATO's over-dutiful attention, Gail, Peggy and I devised a booby trap. With our dust mop we thoroughly cleaned under the beds and in any long-forgotten spot. We had room inspections so often that we probably needed to raid other rooms to fill the mop with silvery gobs of dust. We propped the mop loaded with dust balls against the juncture of door and jamb, mop head up, just at taps. This vigilant ATO was always there immediately after the final trumpet note sounded. That night, she threw open the door hoping to catch us up out of bed as an infraction of the regulations and was the unlucky recipient of a face, head and dark blue uniform full of dust.

Ten minutes after taps our squadron cadet DIs arrived.

"Lights out time, ladies," they chanted with no threat of demerits. "Good luck on your exam tomorrow, Utley," one said, closing the door.

The mop was put away and dust was long gone. Long gone on the dark blue uniform of the unfortunate and misinformed ATO.

12

Hell Week

> "If you're going through hell, keep going."
> —Winston Churchill

We were caught with our pants down—not literally but almost. Our awareness dawned when Peggy went out into the hall to go to the latrine. Gone for just a few minutes, she burst back through the doorway, her face a gray color matching the taut wool blankets on our beds. Her voice seemed to have stayed out in the hall.

"They're gone," she gasped.

"Who? What?" Gail and I spouted, almost in unison.

"Everyone. They're gone. Hell Week has started and we missed it."

OH NO. *This was the worst. How could they? How did we…? What is going on?*

The benefits of living adjacent to the squadron and all these months of becoming one with the upper- and underclass men of Black Panther Twenty-Nine instantly evaporated like alcohol on a hot sidewalk. We were sunk.

"We couldn't have missed it. They would have called us. How could they forget us? Surely, they realized we were missing. We are the only women in the whole squadron," I said.

Our classmates and the other men of Twenty-Nine, they left us? We worked so hard all semester to be part of the squadron. Our skits, our participation in intramurals, sitting at the tables with them for so many meals and reciting knowledge in the halls. Then did they forget we exist on the brink of our acceptance to the wing? There were times during the year the ATOs escorted us to required duties. The ATOs now were more advisory, but their presence in the dorms, at our doors and in formations added confusion about who were directly responsible for our training. *I acknowledge the potential for misunderstanding, but they should have known. YOU, CADETS! WE ARE YOUR DOOLIES!*

But no one had called us. The whole shebang to end this awful Doolie year had begun, and we were here alone in our room. We must have been the only Doolies left, in any room, in either of the dorms, in the whole academy.

"Terrible. Worst thing yet. What in the heck are we going to do about it?" moaned Peggy.

"There's nothing else to do but get ourselves over to the assembly in Arnold Hall. Maybe no one will see us come in," my pure pipe dream shot out of my mouth.

We had known it was coming. We really had no excuse this time. Finals were over for second semester, and this one last hurdle remained of Doolie year. Hell Week. Hell Week, which upperclassmen talked about with both awe and sinister gleams in their eyes. Hell Week, the culmination of the entire year of training, and it was to be three days (originally a week before our time) of nothing but the Hades-like fires of torment. It did, though, hold salvation at the end. The Three Degrees would spend these days screaming away their voices, re-asking us all the knowledge in *Contrails,* and running us everywhere we went again. It would be like all the worst times of the entire year including Basic, pressed upon us in three days. It was our final test. When we finished, we would finally, with all the ensuing privileges, become one of them. We would be upperclassmen.

The problem was, we had grown warm with the complacency of belonging and forgotten to stay vigilant enough to double-check when it actually began. Upperclassmen confiscated our watches and clocks earlier in the day to set us up for the shock of the kickoff. No underclassmen were supposed to be ready for Hell Week. It was just supposed to come crashing down, avalanche style, to bury you or drag you along. So we relinquished our timepieces and sat in our rooms to wait. Knowing we were but a privacy door away from the men in our squadron, we sat on our beds at the ready. When nothing happened right away, we relaxed.

"It's starting," Gail had noted once during our vigil. But it was distant shouting. We rationalized that it was coming from some other squadron or another section of the ladies' area. No one banged on our door. We let our muscles relax again and our pulses slow.

What we didn't hear, didn't know, until Peggy saw the evidence, was that the privacy doors had been closed. The sounds we heard, all muffled through the heavy doors, *were* the sounds of our classmates being jolted out of their rooms. We did not hear them being herded up the stairwell on the far side of the squadron. We missed the cacophony of yelling as the entire cadet wing began the run over to Arnold Hall amid the screaming of 1,000, 2,000 upperclassmen for the start of this trial. We now heard the silence of being left behind.

12. Hell Week

"I can't believe they are gone. They didn't call us. They didn't even think to come and get us. We have been a part of this squadron for five months, and today they forgot we exist. What about all the other women?" I said, both anxious and annoyed.

"They went to their squadron's air power rooms to wait." Peggy knew. She had talked to Diane and Sandy across the hall. But we had been smug. *We don't have to go anywhere. We can sit right here like the Doolie men and wait in our own room. We belong right here in Twenty-Nine.*

Yet, it did not work out that way. We were the ones who not only didn't belong. We were the only ones out of the cadet wing not participating.

Gail's voice reasoned, "We have to go. The more we miss, the tougher it will be." She knew it was another rite of passage, dreaded but required. "We have no options," she added.

On her sage advice, we sprinted up to the terrazzo and along the deserted marble strips toward Arnold Hall. It was eerily quiet, unlike any other time I had been on the terrazzo. It was deserted too. Even during the historic snowstorm, the whiteness blanked my view to mask what might have been others in the same scene. Now we could verify that there was no one else sharing our misfortune. In our tiny column of three, I stared stone-faced at the back of Peggy ahead of me. She likewise focused on Gail's back, and we ran.

Peggy's blue shirt suddenly came closer, almost touching my nose. I skidded, pushing my low quarters shoes hard against the slick white marble. *What's—* But looking up, I knew. I dreaded the sight.

"Good afternoon, sir, Black Panthers," we quietly said together.

"You three are going to be three fried Black Panthers. You are missing Hell Week," he said, commenting on the obvious.

"Yes, sir," we all sighed, expecting the hell to begin with him.

But he just smiled, a slow, kind, sympathetic smile. He knew what we faced and knew he couldn't help us. "Good luck, ladies. You'll need it." And he moved away, rattling his head as if to shake away the idea of our awaiting trouble.

He was of a different mindset to let us off like that. No surprise, really. Any cadet who would walk out on the opening of Hell Week probably had no love for the screaming that was going on. He would not be interested in screaming out here, one on one, either. There were cadets like that. It was hard to identify them, though. They kept quiet, did not pepper Doolies with knowledge questions, or did not get in Doolies' faces about uniform discrepancies. They stayed to themselves or with their friends. They concentrated on other aspects of cadet life. There were serious athletes or serious academics who were not overly interested in training and harsh military mindsets. In upper-class cadets, too, USAFA was not uniform.

There were distinct individuals and this one gave us a brief respite before the melee. *Thank you,* I whispered to myself, the peaceable cadet and God.

"Why do you think he let us off?" I started, but Peggy ran on, wanting to just forget him and get ready for the blitz.

It really was hellacious. In the large theater of Arnold Hall, the darkness startled me. The beam of light we let in broadcast our lateness. As my eyes adjusted to the dimness, I saw all my classmates standing at their seats in the rows in a stiff attention posture. My ears adjusted to the single voice. The cadet wing commander's piercing words about "the need for your class to take this final challenge" were the only sounds. I lost the rest of his speech as upperclassmen converged from what seemed like 12 different directions to whisper their tirades in our ears. "You stupid ladies are late to your own Hell Week." "I guess you don't want to become upperclassmen." "We really don't want you if you think you are too good to join your class for Hell Week." Trying not to disturb the entire assembly, their whispered denigrations were worse than outright screaming would be. Almost each one promised us special attention later in the evening or during the coming days. They left us so we could join the rest. Shaken by the seeds of threats they planted, my apprehension grew.

From the moment the cadet wing commander dismissed the assembled Doolies, I lost sight of Gail and Peggy in the pool of blue underclassmen. It was Basic Training all over again. Wearing their green fatigue uniforms and USAFA T-shirts, the upperclassmen, specifically the Three Degrees, exploded in screaming commands. These third classmen were explicitly in charge of Hell Week, but Firsties and Two Degrees joined the kickoff. When upperclassmen swarmed in to jump-start these final days of training, the ratio of blue to green overpowered my senses. Every Four Degree was surrounded by upperclassmen. Blue shoulders disappeared behind fatigue green backs. I had no chance of finding my roommates. I'd meet up with them later to compare notes. For now, we were on our own.

"Move out, move out. Get going, you lousy Doolies" brought my attention back to the task at hand, to get out of this building.

"Keep those legs moving. Remember, you are back to *running* when you hit the marble strips, Utley," a 9th Squadron Two Degree shouted. Gail, Peggy and I were unlucky now to have two sets of squadron upperclassmen feeling responsible for our training. Niners and the Twenty-Niners knew us by name and by weaknesses after wearing their squadron patches.

My thoughts shrieked back at him, but my lips were tight against the thoughts. *Three days. I can make it.* "Yes, sir," I replied.

I ran along the marble strips through the gauntlet of hovering upperclassmen. Like bears at the edge of a salmon-thick river, shoulder to shoulder, the upperclassmen squeezed against each other to have a front-row

12. Hell Week

place by our path. I welcomed the chapel wall against my right shoulder as a one-sided barrier to the harassment. The solid line of upperclassmen to my left was obstacle enough. Running behind a classmate, I chanced to escape as a raging trainer "pulled out" a cadet in front. But there was no total escape. The trip to the dorms was a leapfrog of escaping one "training session" only to be caught in the very next trainer's claws.

"Drive out here, Utley." My body stiffened. I made the requested left face and took one step forward to release the flow of other Doolies behind me for the awaiting trainers "downstream."

They demanded knowledge answers. The next trainer delivered a nose-to-nose reprimand for breathing without permission. "You still don't belong here, Utley" or "I am going to make sure you don't make it through Hell Week, Utley. You are never going to be an upperclassman at my academy." Miraculously, sometimes a whispered word of encouragement followed. The line and the "driving out" continued even through the stairwell and halls of the dorms as each upperclassman had a chance to add their corrections and comments. Stopping so often, the five-minute distance held us for 30 minutes. I neared what I thought was the end of the gauntlet but never quite reached the safety of the inside of my room. Louder than the 0545 awakening of Basic Training days or the outdoor shouts of assault course cadre, upperclassmen herded all 29th Squadron Doolies into the hall of the squadron for a crescendo of training with all those voices confined to the enclosed hall space, filling every square inch of it. It seemed as if the upperclassmen had a quota of decibels to use up before the end of the year. Hell Week was the last opportunity to expend their ration.

Gail, Peggy and I hardly saw each other during the next three days. Special sessions of yelling and push-ups and strangely constructed obstacle courses interrupted even the nights. Fists battering doors again replaced alarm clocks.

Upperclassmen demanded a year's worth of *Contrails* Air Force Days, aircraft inventory and memorized quotes. Again, I had to recite the discipline quote or the entire transport aircraft inventory. "Yes, sir. The transport aircraft inventory included the Lockheed C-5 Galaxy, the McDonald Douglas C-9 Nightingale, the Lockheed C-130 Hercules, the Lockheed C-141 Stargazer...." I retrieved some of the early knowledge from dusty cranial recesses, unused and hidden behind the English, calculus and engineering facts of the year. We punished our legs, running again everywhere we went. We ran in place in the halls holding our rifles in our arms.

We changed from one uniform combination to another like chameleons changing colors crossing a patchwork quilt. Susie's hair routine would never have made it. There was no time for curlers with inspections in any and all uniforms at any time, day or night. Gail, Peggy and I were now lucky in our placement in the room right next to the squadron. A ten-minute change of uniform worked for us because our room was just one minute from the men's. The other women in both women's areas, the original still in Vandenberg Hall area and the one in Sijan Hall, had to run back and forth from their rooms to their squadrons, attempting but not fully able to stay in sync with the schedule that their male squadron mates managed. Many of the other women surely ran many more steps than we did and probably had much less time to eat during the hellacious time.

There was word that leadership considered ending Hell Week early as they watched from the windows of their offices and from vantage points around the terrazzo as the women seemed to be constantly running the marble strips to and from their squadron and to and from Mitchell Hall for meals. Thankfully, they didn't stop it all because we needed the traditional closure to the year and we didn't need any other changes that would be blamed on us.

There was no escape, and although every fiber of every muscle ached, I did not want escape. This harassment and trial were the culmination of our Doolie year. We were thousands of paces and maybe hundreds of push-ups but only days away from the goal of being upperclassmen. I would have endured almost anything.

Basic Training was so different because we had not yet realized that the goal was to indoctrinate us as part of the cadet wing. We rarely ever learned the reasons for what we were taught at the time. We blindly obeyed and endured, waiting for the end.

Hell Week was different. I knew the goal, and it had become mine. I had earned a small foothold in the cadet wing and I wanted to continue up the ladder as an upperclassman. Not just for the privilege of leaving the marble strips and walking through the middle of the terrazzo. Not just to be able to eat in Mitchell Hall one chair closer to the head of the table. Not just to be on the other side of the training: the trainer not the trainee. No. I wanted to be a United States Air Force Academy cadet, a third-class cadet, then a second-class cadet and, eventually, a Firstie. I wanted to graduate and become part of the "real Air Force" of cadet dreams. I had become one of them in my mind. Now I just needed the tiny silver prop and wings of upperclassmen as proof that I belonged.

At the final ceremony, we Doolies gathered on the terrazzo in our squadron formations. The last screaming had accompanied us up the stairs. A hush fell over the terrazzo at the cadet wing commander's directive to "fall in." A peculiar stillness filled the delay between commands. This window of peace was my moment of reflection and an opportunity for the goose bumps to rise on my arms and tingle my neck. The commander continued, "Classes of 1977, 1978 and 1979, fall out. You may now recognize the cadets of the United States Air Force Academy's Class of 1980 as upperclass members of the cadet wing."

My Hell Week trainer moved in front of me for a final word. He whispered his remarks to me. I pressed the toes of my shiny black low quarters hard into the terrazzo. Still stiff at attention, I strained to catch his words.

"This is it, Utley. It is all over, and you made it," he said.

His hand drew a tiny silver propeller crossed by tiny silver wings out of his pocket. "May I touch you?"

"Yes, sir," I said as I stood at my proudest, tallest attention.

He pulled the lapel of my Alpha jacket up and stuck the prongs of the pin through the fabric. Attaching the brass caps over the back, he straightened the lapel and took a step backward.

"Congratulations, Kathy," he said in a modulated tone. He put out his right hand.

The "Kathy" sounded foreign. I was a human being in name now. I could call him by his first name as well.

My brain just wasn't ready for that adjustment. "Yes, sir. Thank you," I said.

I took his hand and accepted his handshake, his congratulations and his recognition of me as an upperclassman of the cadet wing of the United States Air Force Academy.

13

American Pig

> American fighting man code of conduct:
> I am an American fighting man. I serve in the forces which guard my country and our way of life. I am prepared to give my life in their defense.
> If I am captured, I will continue to resist by all means available. I will make every effort to escape and aid others to escape. I will accept neither parole nor special favors from the enemy.
> —*Contrails* 1976

I stood in the open shower stall hoping the scalding water would wash away more than the dirt from my weakened body. I wanted the sweat and black shoe polish mixed with sweat that melted into the pores over my nose and under my eyes to flow down the drain as well. I wanted to see some freckles again, although I had mostly hated them in any mirror I ever passed. Mouth open, I let the water pound onto my teeth. I wanted the scum from the rabbit meat rinsed off with the memory of killing it swirled down the metal grate in the floor. I held my stiffened fingers up to the jet of water to beat the dried calf's blood out from under the nubs of my nails. My legs and arms ached from the trek and the pack that I had carried. Let the beating water soothe my muscles like a luxurious massage. But it would take more hot water than this shower could deliver to wash away the remnants of survival training that dirtied and challenged me during the past three weeks.

In May 1977, I finished the long and arduous first Doolie year. Then, after the fires of Hell Week, I was an upperclassman. The summer awaited but not with the freedom afforded college students who pack up, return home and live with parents to be coddled a bit and spend time with friends. Academy cadet summers were divided into three three-week periods, only one allotted for leave, going home. The curriculum reserved the two others for academy summer programs. During this first upper-class summer, we chose one elective program, but one was already chosen for us. SERE was

13. American Pig

the required summer program for all the rising Three Degrees. It was not at all the mental picture that forms with the words "summer program"; it was nothing like summer camp.

SERE stood for Survival Evasion Resistance and Escape, only there was no escape and no evading the process either. It was the training for potential survival after bailing out of an aircraft and the preparation and training for a possible prisoner of war experience. An absolute graduation requirement, it would have been chosen by few had it been an elective. Other summer programs, soaring or free-fall parachuting, were sought after and awarded as prizes for high military standing or an attaboy for a job well done during the year. Not SERE. SERE was like an inoculation, a required painful process that immunized for a future that might never even include any exposure.

SERE was divided into three parts. The first part was classroom training where we learned about how to travel along contour lines depicted on a map when evading capture so as never to be exposed on the crest of a hill. We learned which berries were edible and which could cause severe stomach cramps and diarrhea. We learned the POW tap codes so if ever captured we could communicate with fellow prisoners. We heard true stories of brave POWs, some of whom were academic instructors at USAFA.

The rest of SERE was for practice of what we learned in the classroom. One part would be exercises in survival and evasion. The third would be a prisoner phase for training in resistance and escape.

We could have made a case for the women not to be included in the torture. At that time, there were no combat positions open to women, therefore, no need for women to complete combat-required survival training. However, there was no way, after all we had endured to become a part of our class, we would not have fulfilled all the same requirements as the men of the cadet wing. As it was, upperclassmen had nicknamed the Class of '80 the "Burger King" class. According to them, we were the ones who supposedly "had it your way" during the Basic Training summer of the previous year. Any changes, no matter how inevitable, were attributed to our weakness as a class and, overtly or covertly, eventually led back to the inclusion of women. So if SERE was on the list, we would experience SERE and be glad that we did.

After it all started, I would have eagerly opted out.

* * *

As a diary of the survival and evasion training ordeal, I wrote directly on the back of the three-foot square topography map that we used to navigate in the woods. Written in black ink and dated 14 July 1977, 0730, the story was broken up by creases of the folds that shrank the map to a manageable six-by-eight-inch rectangle that would fit in the makeshift backpack that we carried. If I had been found with this map marked, and therefore compromised, I would have immediately failed the whole SERE program.

I think I'll start to write now while it's all so fresh in my mind. SERE has probably been the most difficult, mentally and physically, task I have ever undertaken. It wasn't, of course, voluntary; it is a graduation requirement. I was scared when it began, and I felt like I hated every minute as I went through it. Then I ask myself, "Why are you writing it down? Do you really want to remember it?" I do!! Not because I liked it, but because now as I sit here 100 yards from my last checkpoint, I have a feeling of self-satisfaction like I've never experienced before and that I never want to forget.

Survival, the first section of training, was not first for me at all. It wasn't until the last week of this three-week program that I began the actual survival stage. It was all really a challenge to see who would survive, but since each stage has its own name, I'll try to keep this orderly.

Survival began with two days of "classes." We hiked around the mountains in elements of 6 cadets with our survival instructor to practice using maps and compasses and to visit the demonstration area where we saw 101 things to make with a parachute. Most of them seemed pretty ridiculous, but I guess if you were lost and had a few weeks on your hand, and, of course, the enemy wasn't chasing you, some of them might be fun to try. In the afternoons we made our own little ingenious inventions like canteen holders, backpack straps, sleeping bags and fishing nets. Of course, these too were all made out of parachutes. One afternoon, as I sat sewing on one of my creations, I wondered what in the world would happen to an Army guy if he got separated from his company? I mean, how could he survive without a parachute? I guess that's the breaks.

The third morning we packed up all our gear, strapped it on our backs and kind of stumbled down to the buses. I wondered then how I was going to manage 10 miles up and down hills when I could barely get down six flights of stairs. The bus ride was two and one-half hours. Everyone around me was stuffing his face with candy because we all knew the great famine began the minute we got off the buses. One week in the woods without all the luxuries. I didn't believe in the stuff method. I figured that the less I ate beforehand, the smaller my stomach would shrink, and I would feel less hungry. I am not sure if it actually worked or if I wanted it to so badly that I just kept telling myself that I wasn't hungry.

After the buses, we hiked to an area covered with a parachute canopy that was going to be my element's static camp for three days. After depositing the packs on the ground, we each began to build our first parachute shelter. It was an A-frame type structure with the chute material stretched and

13. American Pig   165

staked over 3 poles. Just as we were finishing, it began to hail. That almost ended it for me. It was bad enough that we were going to have to sleep in the shelter I made, but if it was going to hail.... I was ready to go home. After about two hours it finally stopped, and our instructor, Sam, went to pick up the rabbit that was going to be dinner.

The guys were starving by now. Two of them weighed over 200 pounds and they were so desperate that they began chasing squirrels. I wasn't really hungry. Mostly, I was depressed.

I had camped growing up and was not daunted by the outdoors. I had been home for three weeks at the beginning of the summer and felt rested from Hell Week and the remnants of Doolie year stress. But I had finished the POW experience just the week before this and the bruises on my face had barely faded. My psyche was still bruised, and I was tired of proving myself to all those monitoring the successes and failure of the women.

To top it off, my team decided I had to kill the rabbit. When Sam brought the cute little white thing back, I broke. Maybe if I was really hungry, I could do it, but not then. He agreed to let me off the hook, but surprisingly all the other hungry guys wouldn't do it either. When I realized that they were just as afraid as I was, I regained my courage and clubbed it a good one at the base of the head and it was all over. The feet were given to me, as I was the assassin, and Sam kept the skin. Its meat, one potato and one-half onion made dinner that night and we all hit the sack.

Day 2 of static camp had to be the worst, mostly because it was the longest. Up at 6 and tore down the shelters. If I were really in a survival situation, I certainly wouldn't tear down a shelter that took me three hours to build after one night, but that's USAFA. With ponchos, cans, sharp knives and of course, a gore of parachute material we marched out into the Demo Area 1. Various persons showed us how to blow whistles, shine signal mirrors and shoot flares.

We were being taught how to signal for rescue. Flashing the mirror wildly into the clouds, I was ready for the rescue part to coincide with the training. Could we possibly experience rescue without the survival portion?

The cadre shot a calf and everyone had to get their hands bloody carving the meat out. I'm sure the USDA would not have approved of the dirty hands or the fantastic cutting jobs, but by dinnertime none of us cared.

A commercial-sized metal coffee can with the warm blood of the calf was passed around the 150 cadets sitting on the matted grass of the field. The cadre told us to drink some, at least a sip. No one said it was required, but no one before me seemed to skip a turn either. I had watched carefully. The warmth of the can surprised me when it reached me. It was still calf body temperature. Using two hands, I raised the large can to my lips and tried to look unconcerned about drinking. I planned to barely sip or not even sip and kept my lips pressed tightly in the front. Almost my whole face was within the circumference. I closed my eyes as I tipped the can up. Because the rim was a bit wider than the can, because of the ridges in the design of the can, and because I had I didn't want to see the liquid as it approached, the warm blood of the calf surprised me, flowing into my mouth through the edges not sealed around the rim. It also splashed over one of the ridges leaving drops on my upper lip. Lowering the can and gladly passing it to the cadet to my right, I swallowed the sip in my mouth and licked the drops off my lip. I don't remember the flavor or much of the smell. It smelled like metal, like the pull-up bar from the gym where I went to get tested for my original appointment. It smelled like the barbed wire from the assault course. It smelled like iron. Maybe that was the can. I did not gag or grimace; others waiting for the can were watching. I was satisfied to have drunk my tiny portion.

I did not make the lasting impression on the entire group as former squadron mate June Van Horn did. She popped the offered eyeball of the calf into her mouth. She held it there like a gumball, ballooning out the skin of her cheek, then, after it slipped through her teeth to the other cheek a few times, popped it. June described it: "I was singled out as one of three lucky folks to eat a cow's eyeball. It was like firm rubber with no taste until you finally bit through it, and then it was just like salt water. But the eye must have been under pressure because when I was finally able to bite a hole, the water just spewed into my mouth and forced the gag reflex. We were all going through this. We already had pictures with smiles and the eyeballs in our mouths. But now, no way were any of us going to lose it. So somehow, we were able to—I won't go into gross details here—but we all finished the task."[1]

At about 14:00 we trotted back to camp with a small can of blood and our share of the meat. We built lean-to shelters and 60 percent of the meat was cut into small strips pounded paper thin and heavily salted to be smoked into jerky that night. Most of it was so full of dirt that I swore I'd never eat it. I was wrong. Another potato, the blood and the rest of the meat made dinner that night. Shifts began to make sure that the fire in the smoking tent didn't go out.

Day 3 was the easiest yet. Tore down #2 shelters and the third were just evasion type which just meant a hole to hide in. I found mine fairly soon and then goofed off. We had a short lesson in triangulation, divided the jerky, had a lesson in camouflage and were finished for the day. Dinner was potato stew, and it was delicious. A couple of my pieces of potato fell on the ground but after wiping most of the dirt off they were o.k. Sleep that night wasn't so good. The hole I'd found wasn't too level and by morning I wasn't much hidden. I had rolled right into the middle of the trail.

13. American Pig

Day 4 was the beginning of the "trek": 4 days wandering in the woods, looking for checkpoints and being chased and shot at by the group called the aggressors. The chaplains held Masses that morning, which really made the beginning seem more dangerous. A small out brief and we were on our way. Checkpoints are only open for certain lengths of time so you can't just stroll to them. We got a card from which upperclassmen playing the parts of aggressors or partisans can take points off for being late, getting lost, getting caught etc.

Getting to checkpoint #1 was easy. Our instructor was still with us, and he practically just led us there.

I called on our instructor to intervene for me with the cadet playing the partisan and probably stretched his tolerance for women in this environment. I started my period that morning, many days before I was expecting it. All during Basic Training, the strain of the environment kept most of the women free from the hassle of menstrual cycles for the entire time. I was thankful, then, that I had no worries. I knew that whatever was keeping my period at bay was a blessing and no cause for worry. Now, starting this trek into another wild unknown, my body rebelled again against me. I asked our instructor to ask the partisans for tampons for me. He was so flabbergasted by the request that he just staggered away. It certainly must have seemed above and beyond his call of duty.

At each checkpoint, one person checks in with the first partisan then everyone sees partisan #2 and #3 to get their cards marked to prove they had made it and to get directions to the next checkpoint.

I checked in with the partisan at #1 and lay on the ground while he role-played in a post–Vietnam-era scenario, as a Southeast Asian man in danger of discovery, begrudgingly willing to lead me to the others, that would send me to another group, that would send me closer to "rescue." His hostility reached its peak when he told me to "move over towards my friends near those trees" and losing some of the accent, he handed me two tampons and said, "Now go."

They split us into two teams of three after #1 and sent us on our way. I was with an ex–Boy Scout and a guy from Colorado Springs. Things seemed to look up. I did not feel very confident with a compass, but with a boy scout and a guy from the area what could go wrong? Well, everything did. First, one took charge and ran us right into some aggressors, which cost us 5 points. Then the other who was supposed to be the brave one turned out to be the biggest chicken I have ever seen. The first time he heard a shot, he took off running, forgot about our course and ended up getting us lost. It was my turn. I just approximated our location, took a heading and stayed on it.

We found a place to sleep by 1830 (have to stop by 1900) and went to sleep. Oh yeah, I ate a cornflake bar for breakfast and the ham slices from my c-rations for dinner.

We carried cans of black shoe polish in our packs. They served two purposes. I blackened my face with the polish to be less visible on the mountainside. It was required; points would be deducted for an unpolished countenance. Gail's face, a beautiful mahogany color naturally, even had to be polished. She argued her case, during an earlier SERE period, to no avail and wore the grease under duress. I wonder if another male cadet with a dark complexion had argued that case before her. We also lit the polish and used it, like Sterno under a chafing dish, to warm the awful contents of our C ration cans.

In the morning, I was voted the leader and we hit a road 100 yds from our checkpoint #2 by 0700. It didn't open until 0900 so we found a hiding place and waited.

By this time, I was desperate for some more tampons. I approached the first partisan myself, forced to have to tell the whole story to another unsympathetic and disgusted '79er that I could not go three days on two tampons. "We are here to see you safely to freedom," he said in a fake Russian accent, staying in character. "These problems of yours are not of our concern." I asked for an "academic situation," our code for a time-out from the role-playing, to get genuine help. The partisan was uninterested. But word had reached the checkpoint ahead of me and he knew what I would ask. "Take these," throwing a small box of ten tampons at me, deducting points from my tally sheet and yelling to "get out of my sight."

I got docked 15 points by the other partisans but since it took 150 to fail, I wasn't worried yet. We got our info, checked in and took off for #3. Since it opened at 1700 on the same day, we didn't fool around. I took the lead again and by 1500, we were there. Found another hideout, ate the peaches from the C-ration for lunch and waited. By 1730 we were through and heading for #4. We didn't go far since we had until 1300 the next day. Slept on the side of a ridge and had ½ a LURP (cooked and dried) package for dinner. I had lost the description of #4 somewhere along the way and thought for sure we were goners. Just then another team came by heading for the same point and let us copy theirs. I felt lucky!

Up again at 6 the next morning and I was leading. Somewhere along the way I lost my canteen cover, so I had no water. One teammate was willing to share. We came out about 200 yards downstream from #4, walked till we found it and waited 2 hours for opening. I had peanut butter and crackers for lunch and just waited it out. I couldn't sleep with the pack on, so I just sat in the bushes and waited. When we finally checked through, I felt relieved. Only one more to go. Leaving #4 wasn't easy. The underbrush was thick, and we knew we had three miles to go. I led again but not very quickly.

13. American Pig

I stopped as I took each heading, and we took 15 min breaks every so often also. We were all pretty dehydrated. I started feeling nauseous and dizzy. We passed one small creek, but it was only mud, so we decided to wait. At the top of one hill, long after I had quit evading, my heading led me to a tree right out in the open. I was too tired and thirsty to care anymore so I just walked over, whipped out my trusty compass and started searching for the next tree. I got pretty good at picking out certain trees. I guess maybe the trees were insulted but it kept us on course. I'd just sight on one, try to get a mental picture of it then pick out a distinguishing feature to remember. Sometimes it was, "that bony one" or "the one with the two knots right near each other" but with all due apologies to the trees it got us there.

We finally found a large creek, filled with beaver ponds, and filled up, waited the 15 minutes for the iodine tablets to work, drank, then filled up again and started on our way. By 1815, we were thoroughly beat. We forgot all about hiding and slept out in the open on the side of the ridge.

I was fairly sure where we were, but a road that I had expected, hadn't shown up, I could still see the beaver ponds. I took a heading, figured our position and sure enough the road was 100 yards behind us.

I remember gunshots and lots of whooping and hollering that night but was so exhausted that I could not rouse myself enough to even think clearly. I wonder now if the "aggressors" were out to exact the last few points from the SERE evaders or if they were off duty and just having a good time on their last night out in the woods. I understood from classmates who "worked" SERE as upperclassmen that it was a great time. While we ate our freeze-dried foods, our slimy spaghetti from its three-year-old can or our jerky sprinkled with dirt and smoked with the aroma of parachute nylon, they were eating steaks and chowing down on Oreos.

Sleeping that night was really rough. My spot ended up being more slanted than I expected. I had dreams of choking and had to keep getting up to get water. I kept my boots in the bottom of my sleeping bag at night [to keep spiders out of them and to keep them dry] so to top it off, every time I wanted to switch position, I had to fling the bottom of the bag, weighted down with the boots, around first then follow with the rest of my body. I was really glad to see the sun come up.

We took our time getting ready then went straight up to the road and followed it right to the checkpoint. Once again, it was marked with parachute material, but I wouldn't have cared if there was my favorite dress hanging from the trees, I still would have been elated. We hid among some fallen pines (there had been a tornado recently) and I began writing this to keep myself occupied.

I imagine that the guys thought I was nuts. They were probably melted into the dirt in absolute exhaustion knowing that the ordeal was a crawl from being over. I guess they had no yearning to write the whole adventure down or get in touch with their emotions about its completion.

Writing helped some, but just knowing that that one was the last made the minutes drag into hours. 1000 finally arrived. It was one cadet's turn to go first. He hadn't wanted to lead that morning when I took a break for a while, so at least he could check in with the first partisan. He did when he sensed that we were serious, and our long-awaited escape began. We went to the second partisan, as usual, but they didn't just mark our cards and go on. We laid face in the dirt for a while then they ran us to another spot for another face down rest in the dirt. We filled out our cards, with fingerprints even. Next, we were taken down a little valley and picked up a piece of fruit. I got an apple and it really tasted good. They were talking to us in accents and calling us pigs and stupid animals, but it wasn't like the compound. I knew it was almost over and I really couldn't pretend that they were partisans, and I was a downed officer. Afterwards, it was essentially over. We ate our fruit, waited a few hours and walked to the bus pick up point.

It is now 1350 and I'm finished. The buses will get here sooner or later. Box lunches, and a shower. I can't believe it.

The trek portion of SERE was over!

* * *

Part one of the SERE experience was not as outdoorsy as the trek portion. The shower after that 48-hour training session was short and careful. I wasn't that dirty. My body was bruised then, and I needed sleep. A short wash, a long rest, and then I called home. My mother answered. I was so glad that there were no visuals in those days accompanying the phone calling. My face was bruised from ear to ear along my chin line from the beatings. My mother didn't need to see that. But when Dad got on the line, my voice cracked and the line between my receiver in Colorado Springs and his in Florida was quiet. I couldn't talk. I didn't want to cry. I didn't want him to have to talk in front of Mom about it. But he knew. "It was tough, wasn't it?" he whispered. I nodded, and even over the phone he knew. "I knew you could make it. You are one tough young lady. Relax now. It's all over." He had never been a POW, but he had captured some back in World War II. What he knew was beyond what I knew from my 48-hour experience but no longer beyond what I could imagine.

I also kept some notes during the SERE prisoner of war simulation. It was 48 hours in a barbed wire–surrounded camp. It was an intense re-creation of a prisoner experience, physically and mentally, with no breaks except for emergency "academic situations." We were strip searched after "capture," but I had sewn paper and a small pencil into the hem of my fatigues. I had some Tic Tacs in the waistband of my fatigue pants, too. My son, who many years later went through SERE, laughed when I suggested he do the same for his 2000s-era survival simulation. "I don't even like Tic Tacs," he said. Me neither, but it was not about taste.

During the week before, we knew capture was coming but not the date or the hour. We walked around, back and forth to SERE classes and training, sometimes noticing that other groups of cadets disappeared for a few days. I knew our turn was coming. The paper and tiny pencil and the candies were my minuscule act of defiance during the anxious waiting times. The paper was small, only three inches by five, folded into the size of a stick of gum to fit in the fabric. There was no room for excess adjectives or prepositions. But after the words marked its surface, they would forever define a new part of me.

1600 captured in quad. Bags over heads. Cattle packed into truck and driven away. No turi [tourists] see.

I knew we were in a blue government truck, and I could feel the bodies of my classmates pressed against me. My joints were stiff from the cramped position and the bouncing of the vehicle. The upperclassmen were fully in their roles as our captors with foreign accents and rough pushing. The green canvas laundry bag on my head smelled moldy and also had a weird vinegary smell. If I hunched my shoulders forward, I could see the floor of the truck and some bit of light. I could see my own combat boots and some of the boots of the men around me. I couldn't see any other women's boots. Somewhere there were about ten other women in this flight of prisoners.

Face down in dirt, ants bite arm.

I never knew if I was just so unlucky that when the "camp guards" yelled, "You, American pig, get on ground," I ended up on top of an ant pile, or if they made everyone lie on that same pile. I felt the ants first on my hands. I moved a bit and shook them off, but the guard held his boot on my wrist so there was no more moving. The ants crawled in through the loop above the cuff of my fatigue shirt and took a few bites.

When my siblings and I were children at Mass, my mother used to make us kneel up perfectly straight. No touching the seat of the pew behind or leaning into the wooden backing of the seat in front allowed. We competed silently to see who could balance the longest. I think we were supposed to be offering up our strain to Jesus for all the real suffering of the world. In the SERE camp, in the pile of ants, I offered up the bites and pretended to compete with my brother Pat to see who could lie in ants the longest and not move.

Short ruff interrogation almost cry. Strip (obscene remarks)

The camp personnel's harsh questioning and rough pushing was way beyond what we heard or felt in Basic. There were no television crews here watching, but I assumed some officers were keeping surveillance. Probably. Red stars and sickles adorned the uniforms of the guards. I knew they were upperclassmen but did not recognize any faces.

The guards told a group of women to take off all our clothes for a medical exam. "Put the bags over the ugly faces," they screamed. Then I heard jeering and crude remarks about those scrawny ugly American bodies. I assumed the taunters were somewhere outside the tent where we had stripped. They had to be. I trusted they were. Many years later, a male classmate smiled in relief when I recounted the doubt that the upperclassmen as guards could see us naked. "I'm so glad to hear that you stripped and stood like that," he shared. "I always thought that I was the only one that had to do that. I was embarrassed and never had the nerve to ask anyone about it."

Big black isolation boxes. Unsuccessful tap code. Too many at one time, so yelled.

The big black box was not big. I could not stand in it, and I could touch both walls without completely straightening my arms. It was relief from the face-to-face harassment, but it was hot.

Other classmates were "prisoners" in adjacent black boxes. We had learned in our SERE class time how prisoners in the "Hanoi Hilton" prison in Vietnam had communicated using tap codes on the walls of their cells. The code was introduced in June 1965 by POWs held there: Captain Carlyle "Smitty" Harris and Lieutenant Commander Robert Shumaker. It was based on a grid of letters with taps in pairs for the coordinates of the letters on the grid.[2] We were all eager to try it and defy the guards with our secret communiqués. The taps were muffled and slow. Cadets in boxes behind and beside were tapping all at once. We finally gave up and pressing our mouths to the seams of the boxes just "talked" to each other. The guards slammed fists or sticks against the outside walls and demanded we "shut up, American pigs." I knew they wouldn't shoot us. The Hanoi Hilton "guests" had had no assurances.

One legged stool, field jacket, fraction of a blanket, "pee" can and cup. Could see out. Saw houses, guard tower fence, etc. No food. Water every hour. Checked every 15 min. Long time.

The "people's stool" had only one leg nailed through the middle from the small wooden top. Luckily, my butt fit on the top, and with my legs spread out to make the other two balancing points of the stool, I could lean back with some equilibrium. The larger cadets must have sat on the floor. I remember being able to see a small portion of the camp through a crack that ran alongside the door. Did we spend the night in them? A coffee can in the back was the latrine.

June, who was one of the other women with me in the compound, remembers, "In the compound, one of the 'torture' devices was to squeeze us into this skinny structure made of 2 × 4s. The poor football players—I heard some of them were claustrophobic.... The weirdest part was hallucinating and being aware of it. We were in these little isolation rooms with a can for toilet needs and a stool that wouldn't balance. I tried to sleep by standing up and leaning my head against the door. The track under the door was a regular run for ants. But after countless sleepless hours, the ants became Indy 500 cars and there was a regular race on!"

Removed to smaller box.

The small box was very small. Could they all have been the same size? Could big, tall classmate George Madson ever have fit into a box that I felt I could not breathe in? My knees touched my face. My shoulders caught on the sides, like I was a bagel too wide for the toaster slot. Sometimes it felt good to be snug and I would hang by my shoulders to give my knees a rest from holding me up. Little bodies and tiny frames drew the longest straws during the little box time. Female advantage. Cramped and in total darkness, it was almost impossible to sleep, but I dozed or imagined that I did.

Then to interrogation. 1st friendly then more angry. I first Jane Wayne later only babble. Born Miami, U of Miami, ROTC, Chauncey AFB. Parents' names (they) already knew [like a fist in the gut when they said Harold and Mary] Stress positions. Signed receipt too far down.

The interrogators used the good cop–bad cop routine. We had been taught some techniques for disrupting interrogations and stalling during our SERE classroom time. John Wayning was acting like a tough guy and giving them nothing, but my nerves couldn't match the Duke's. Babbling, a stall technique, turned out to be more my natural style. I lied about how I was born in Miami and "some of the people actually wore flip-flops every day, and how right down the street from my house was a stoplight, but you might not even know what a stoplight is but it's easy to imagine, just a light, well really three lights in a metal casing, the top one was red or is it green, no the top is red and the middle is yellow and that means caution, and the bottom one is green for go." "Shut up, you stupid American whore. Answer my question. What was your mission?" "Oh, is that what you asked me? I'm sorry. I was just trying to explain where I come from so you could better understand my background and why I joined the Air Force to begin with ... and how I went to the University of Miami and...."

"Shut up! NOW."

They made us hold our bodies in stress positions until our muscles quivered and then cramped. "Assume the position" started my heart racing. But when the interrogator, claiming to know everything about me anyway, named my parents, my techniques deserted. The shield that I built up between my heart and this game crumbled. Dad and Mom's names broke though the barrier and touched my feelings. I was sad to have my parents' names be part of the ugliness. My lies evaporated. I couldn't think. I forgot all my babbling lies. I signed what they told me to sign but defiantly not on the signature line. With my left hand. Ha! I grabbed any thread of dignity and noncooperation to hold on to.

Almost tricked with Carter quote. Hit a few times. Threatened to take others to box.

The cadre often chose the smallest from each tent as the scapegoats for the hitting "torture" to make us sign a confession of war crimes or an admission of our country's faults. The hitting was not direct punches but a body shaking that rattled my teeth. Cadre grabbed a handful of collar and rolled their fists tight against my collarbone. Like doing bicep curls, they lifted me off the dirt and held my face close to theirs for some direct eye-to-eye screaming. They shook my whole body back and forth like shaking maracas, and their fists banged my chest and my jaw over and over. It was training to imagine being a prisoner of war. I just imagined how in 48 hours it would all be over, and I would hate this face in front of me for a long time to come. I could only begin to imagine the fears of a true POW, fears I didn't feel, of rape, or death, or beatings that would take me to the edge of life. I barely tasted their experience.

Back to isolation then to large tent. Cleaned floor with bags for head and everything dusty. Had meetings and talks about Red Cross. Had election for June [girl chosen as tent leader]. Was seen talking, face in dirt. Work on garden and moving rocks. Guys dug and raked. 9 escape attempts last night only 7 successful. One girl with chain. Called whores. "Spread legs, etc...." Didn't upset me much. Saw Camp commandant. Pictures of Napalm. Pretty bad.

More work and 1 more interrogation. Tried rescue technique and more beating. I continued stalling and babbling. Little black box seemed bigger but hot with field jacket at about noon. Returned to isolation.

Later back to work area. More garden work. Wanted renunciation of Geneva signed, June refused (and they) beat me and Fran until convinced. Sent to area to admit I was a war criminal. Never did. Suffered for other girl instead. Day took forever. No idea of time. Worked more.

Indoctrination from Palestinian. Impossible to win. He told Guards we were "arrogant." Bad news. Antichristianity speech. 207 to wear uniform.

We had numbers written on the backs of our hands after the initial capture. The guard used permanent marker and numbered us in alphabetical order. The women were in a sequence of their own. Andrea Ungashik was 207. I was 208 and June Van Horn was 209. June, 209, was the leader of the women's group. Once in an outside formation, the meanest guard called for prisoner number 207 to come up and wear the uniform of camp

13. American Pig

whore. I stepped forward, confused, tired and hungry. I saw Andrea moving toward the guard, head held high. I looked at the back of my hand: 208. Relief, embarrassment, shame and empathy ran through me head to toe. I stepped back. Andrea was afraid of creeping things. Who knows how they found out? She also had to carry the carcass of a lobster fished out of the watery dinner broth around with her all day.

No God Why not strike rock. He was our God. Made cross and carried tried to burn but rained because everyone prayed.

And as one male classmate carried a wooden cross around the camp, he too held his head high. How proud I was to be a part of this amazing group of young men and women. Each seemed stronger, more courageous and more honorable than the next. He planted the cross and the guards set fire to it. The skies opened up with rain. I knew it was prayers answered.

Worked more. Very tired, sleeping on feet. "No sleep, little one." Meal formation. Everyone collected own. Broth terrible. Drank too much water. Supposedly ⅓ can and tent drank 1. More punishment. Break for rain. 40 min of sleep. Last interrogation. No little black box. Very rough. Wanted to know camp organization. Tried to stall but beaten badly. Needed to pee. When finished very relieved.

Back to big box. Slept. More water. Can full of pee and guard wouldn't empty—finally peed. Box got hot. Anxious for end. Imagined neighbors gone. Completely alone. When [other] answered, thought was fake. weird Music sounded like English. [hallucinations]

Taken to formation. "more work to do." Will face my flag, best flag, flag of freedom, etc., (Thinking we were going to see the enemy flag again).... The American. (they showed us the American flag and it was over)

The end of the time in the compound is fuzzy for me. I was tired and hallucinating. Guards were shaking me to wake up as I stood in the compound almost falling over with exhaustion. We got in trouble as a tent for drinking too much water and I was "punished" with more shaking. Back in the black box, reality escaped me. I was calling for the guard to come and empty the coffee can in the back of the box that we peed in. He laughed and walked away. "Stupid pig," he muttered. Then I thought I was alone, that all the others were gone, not believing the words from my classmates in the neighboring boxes. I couldn't believe anything that I was hearing. Foreign music blasted from speakers sounded like English.

The final assembly of all the "prisoners" in the dirt yard of the camp was supposedly to do more work—hole digging and rock moving. The camp commandant yelled in his accented role to face his flag, the greatest flag ever to fly, the flag of true freedom. We turned and saw the American flag. And it was over.

14

Three Degree

"To fly we have to have resistance."
—Maya Lin

A crowd of cadets filled a doorway down the hall. *Something is happening down there. Trouble? Fun? Training starting already?* Moving into a new squadron after the summer of SERE, my mind still bruised from the memory of it, I was not quite ready to start with the regimen. I preferred to get into my room, settle in a bit, and brace myself for another year, another day. The worst seemed to be behind me, but I had innocently formed that same thought last spring after Hell Week and again after SERE training while the bruises faded. *I am not letting my guard down too soon this time.*

The crowd was at my door, 6F15, Vandenberg Hall. The "6" was for the sixth floor. That digit was not good for me. My knees and thighs wanted to be right on the terrazzo, technically level four, near Fairchild Hall and my comfort zone of the academic buildings. No stairs ever. Upper class now or not, my legs didn't like the workout that even daily life would require. As a Doolie, I had walked up and down from the sixth floor too many times a day. Down for formation to breakfast. Up after. Down for morning classes. Up after. Down for lunch formation. Up after. Down for afternoon classes. Then up again. Down and up for intramurals or drill practice. Down for evening meal. Up after. Add in there a down for a trip to the cadet store, the library, or the chapel and one last up and I had enough ups and downs to equal a wild bungee jump and knees and thighs that screamed for relief.

I would now compensate by using the elevator, one of the privileges that came with the small silver prop and wings of upperclassdom bestowed last spring after Hell Week. "Hold the 'vator," I would call just like I had heard a million times last year from upperclassmen. I could hardly wait to pop in through the sliding doors and push button number six.

The "F" of the 6F15 was great too. "F" meant the squadron was on the north side of the cadet area close to the athletic fields. Athletics were my least favorite activity. I felt enthusiastic to at least be close by, so I didn't have to expend even more energy getting to and from fields that I dreaded.

The "15" was the concession to finally having women among the men. It could have been any number and I would have loved the look of it. That fall, the academy expanded the number of areas for women. No longer just the two women's areas, one in Vandenberg Hall and one in Sijan Hall, now the women would be living clustered adjacent to their squadrons. This Three Degree year, I was assigned as a member of Stalag 17 Squadron, a group of 110 cadets representing all four classes at the academy. As a Three Degree in the squadron, I would finally be living alongside the men of Stalag. Huge. This would make the year a success no matter what else got snafued.

Within the room numbers of 17th Squadron, I really shouldn't have had an odd number for an outside room. They were understood to be reserved for *upper* upperclassmen, Two Degrees or particularly Firsties so they could look out over the parking lots and see their cherished automobiles. But the women had to be grouped somewhere around the converted latrines at the corners of the hallways where two or three squadrons intersected. That way, more women could use one facility. I lucked out on the outside room but felt I won the lottery with the inclusion in the squadron itself. The Doolie women were there too, across the hall on the inside, quadrangle side. Fortunately, they would never know "penthouse" living and segregation.

As I moved through the doorway, I saw a female cadet standing in the room already. Several guys milled around. We were all now assigned to Stalag 17 and would be part of Stalag for our remaining three years. I could live with that.

All the guys in my room turned out to be classmates, Class of '80. I guess the other upperclassmen of Stalag were not anxious to meet the new women in their squadron. Ours was a three-person room with the blue name tags of Burket, L., Hilsman, A., and Utley, K.M., all together. Always aware of inequalities, we checked and saw a few of the male Three Degrees' rooms were triple marked with name tags in the alcoves. We didn't feel singled out. Actually, ours was a preferred outside-facing room. None of the Three Degree men had one of those. Not exactly equal.

Louise Burket greeted me as I walked in. I basically knew her. All the women knew each other. There were so few of us and we had been through so much already that we were the sisters of the brotherhood. I don't think I had talked to Louise much or had a class with her. I had no negative impression of her, though. She was in the 33rd Squadron as a Doolie and had also moved midyear to alleviate the terrazzo gap. I thought we'd get along just fine.

Male cadets were perched around the tiny room. Fred Kornahrens eased out of the room, greeting me with a quiet "hi" as I headed in. Maybe a shy one there. Ron McNeill sat on the vanity with his shoes almost in the sink. His friendly open face shone with welcome. We would be friends. George Madson, Dave Duncan, Bob Schley, Pat Heatherman and Mark Hoffman introduced themselves as fellow Stalagers and classmates. George was a big tower of strength and compassion, Dave seemed focused but fun loving, Bob a brilliant clown full of quips and ideas for mischief. Familiar faces like Pat's from Twenty-Nine and Mark's that I knew from 9th Squadron, bolstered my comfort. Common experiences did too. Recounted stories of SERE and other programs bounced from lips to ears around the room. Did you have first period SERE or second or third? Did the tornado come through your encampment? Did you have rain during the POW camp? How was home? Did you ever think about not coming back? Most agreed that we had torn ourselves away from home and family to return. "If we were Doolies again, I couldn't have done it" started everyone's head nodding up and down. The stories began building the foundation for our class in this squadron: Stalag 17, Class of '80, United States Air Force Academy.

Louise and I bonded. Smart and focused, Louise marched determinedly through the days. Down to her perpetually sharp mechanical pencil and precise handwriting, she managed all her affairs with exactness. She was on the track team, a medium-distance runner and high jumper. Practice kept her out of the room in the afternoons. She ate with the track team, not at squadron tables. Homework filled her time in the room later in the evenings. She was not much of a socializer. She greeted and knew the men of the squadron but stayed focused on her tasks. She seemed an ideal cadet with her strong upright bearing and strict self-imposed regimen of studying and track. I knew I could get along with her because she was so stable. Like a European train, you knew when to expect her, how long she'd stay and how and when she would move to the next stop.

As a Three Degree, I settled in and let the fine starter roots of my comfortable belonging begin to move out through the squadron and the cadet wing. The Class of '80 took on minor leadership responsibilities although only under the watchful eyes of the Classes of '78 and '79. We kept the new Doolies of '81 in line (literally) as we honed their marching skills or corrected them on the terrazzo. They had survived the six weeks of Basic, and now we quizzed them on rote knowledge from the now well-broken spines of their *Contrails*. We slowly proved ourselves as able to train as to be trained. We could lead and follow all toward the end of becoming Air Force officers.

I glided semi-easily along the high-polished floors of the squadron

14. Three Degree

and the academic buildings and even managed the rough fields for intramural soccer and team handball without tripping too often.

Socially, I was dazed but starting to move easier. My cadet summer after SERE ended with three weeks of soaring training to become a glider pilot. I could just begin to think about what happened to my friend and instructor without tensing to hold back the shakes. It had been a tough summer.

Gail and June were my summer roommates during soaring. As upperclassmen, we enjoyed the privileges and the quiet of the academy as cadets only can in the summer. Soaring took us to the hangar at USAFA's airfield every morning and the sleek white swans of the skies. So graceful and quiet, they circled overhead and then landed back on the ground without disturbing the peace of the scenery or the sound of the wind off the range. My instructor was also a quiet gentle soul, and when he was not in the seat behind me talking me through a release from the tow plane or an upward spiral to follow a thermal of air to altitude, we talked. He was a rising Firstie full of hope for his remaining year at the academy and his upcoming years in the Air Force. When the sun dropping behind the mountains pulled the darkness over the airfield, we talked and shared thoughts about USAFA and the Air Force. We went to the laundromat together or to get a bite to eat. We sat on the grassy slope of the south lawn behind the dorms in the evenings and laughed about our crazy cadet foibles. In a short two weeks, we became close friends.

The last Saturday of the program, he was trying for a cross-country badge to sew below the soaring patch that already adorned his athletic jacket. I think the requirement for the patch was a 100-mile flight. That morning, I helped fellow cadets pull the gliders from their hangar and then sat by the edge of the airstrip and watched him perform his preflight checklist. He talked to the tow pilot and gave me a wave, a quick "so long," before his glider pulled by the cable connecting it to the Cessna in front headed off to the south. At the prescribed altitude, he would have pulled the release in the cockpit to sever his connection to the plane. I was following along in my head with the procedures that he had been teaching me so far. I watched him glide a bit. I saw his craft spiral toward the mountains knowing that the vertical velocity indicator and his seat of the pants feeling of lift indicated that a column of hot air was rising from the hot asphalt or some other thermal initiation. He spiraled into that thermal lift gaining more altitude and then banked the sailplane toward the flat lands eastward to get as much distance toward the cross-country patch as he could, spiraling up through more thermals for altitude along the way. Quickly, he was out of range of my eyes, and I stretched out in the grass to read my handbook.

The tweet of a whistle later in the warm windy afternoon surprised me. The noise that we heard on afternoons when rain or thunder threatened came through a perfect blue Colorado sky. "Push in the planes," cadet instructors shouted, and we all grabbed trailing edges of wings to maneuver the gliders back to their hangar.

One of the instructors asked me to wait alone outside while they talked to other cadets. I was confused but should have insisted that I be let in. I sat alone on the step outside the door looking around and up at the crisp cloudless sky. Automatically, my head started praying, but my thoughts drifted from the rote words of prayers learned years ago. *What happened?* bounced back and forth between the *Hail Marys* and *art in heavens. It had to have something to do with my instructor, but why wouldn't they just tell me?* Gail came out the door, and I said to her, "It's him, isn't it? Something happened to him. I can take it, Gail. Is he hurt?" "I'm so sorry, Kathy," she said, and once again she held me close to her heart. "He didn't make it," she said. "No," I moaned. "No," I choked out even louder. "It can't be. It's not true," I growled angrily, and I pushed her away from me. I didn't want a hug. I didn't need it. I could only imagine that she was wrong. They all were wrong. It was a glider, a sleek, silent sailing ship for the perfect blue Colorado skies. It couldn't hurt anyone. It had no motor and no guns. It made no noise. He couldn't be hurt. His heart was too kind for violence, too warm for death. He was my great and caring friend. We grew closer every day in understanding each other's lives and dreams. *You cannot take him,* my scream as silent as the abandoned airfield.

But he was gone forever. Somehow, I ended up in a car and at some officer's house there on base. A chaplain came to counsel me, but I wouldn't hear his words of solace. I wanted to go home. Not home in Florida with my parents but home to the academy. I wanted to be back in the dorms with my roommates. I wanted to be with people who knew me. I wanted my classmates, my roommates, my fellow soaring students who could see me through the grief.

The next few nights in the dorms scared me. Gail and June slept with the lights on for me although sleep evaded me. The cadre of flight instructors and officers grounded me at first, but finally his spirit and a need to finish what I started urged me aloft again. For my final check ride, I flew with his photo in my flight suit pocket. I soared alone in the air as he had before his crash. My hand shook to pull the release cord separating me from the tow pilot's ride. My tense fingers cramped on the stick, and my back hurt, stiff against the seat. Somehow, the mechanics of the flying took my mind from the fright, and I found some thermals and rose farther from the safety of the ground. Spiraling up the channels of air was so peaceful.

14. Three Degree

The views behind the jagged mountains that limited our sight from the ground unveiled bright green patches of grassland watered by melted snow. Far in the distance to the west were more mountains, each ridge farther and lighter gray than our neighboring range. I tried not to look to the east, to where he had flown. On landing, I bounced hard on the tarmac, then softened my body with an all-encompassing sigh to have come back to ground without anything broken. I got my license to fly again. I thought I might, but I never did.

I knew now by the scars on my heart the dangers of flying. The deaths of the Air Force aces and Vietnam heroes that I learned from the pages of *Contrails* were made more real to me by the death of a good friend killed in an aircraft. I still saw swans when the gliders caught my eye in the skies. But I also saw a black floater in the picture, a shadow of the danger of flying that never marred my views before.

* * *

The pain of the summer was diminishing, though, and I had a fresh start in my new squadron, Stalag 17. I met either Bob, Fred or Ron in the halls in the morning to walk over to classes together. The academic schedule was grueling. A cadet designated by the instructor took attendance at the beginning of every class. When the instructor came into class, the designee called the room to attention and announced to the instructor "All present and accounted for, sir" or "Sir, one cadet absent." Every cadet took a heavy dose of core classes. Our curriculum included four years of English, foreign language and engineering including physics, thermodynamics, aeronautical engineering, mechanical engineering, and astronautical engineering. There were "fuzzy studies" too, a nickname for those classes that included more words than numbers and required a lot of reading.

But we didn't study all the time. Fred, Bob, Ron and I hung out in each other's rooms to discuss squadron activities or current events. We rotated duty with other Three Degrees in Seventeen sitting at the squadron charge of quarters desk holding the keys for all the rooms and monitoring the phones. We bonded with the Firsties and Two Degrees as much as they would allow. I could not link myself to the men very much with athletics, but I could and did help with homework. I held my own in the military arena, knowing my knowledge for checking the Doolies but never enthusiastically seeking out more info on planes or battles or heroes.

We played too. We laughed and joked in the halls. We hung out together at squadron parties. We went to Arnold Hall together to get something to eat or for them to have 3.2 beers on the weekends when we had none of our allotted privileges left for a trip downtown. I was in my glory and felt that all the fighting and climbing to find a place for women at this male stronghold had made a difference. We were in. Or so I thought.

There were still two classes ahead of us that were still part of the old guard that had no women in their ranks. With some, we would never be in or ever even be cadets. Some of the men barely tolerated us in their midst and took every opportunity to condemn us for anything. Watching a crime show (*Kojak*, I think) with a mixed-class group in the TV room one night, I solved the case. Before Telly Savalas could get the lollipop out of his mouth and accuse the criminal, I blurted out the seemingly obvious answer. The seating was apartment-type pine-framed chairs and couches with back and bottom cushions of gold and tangerine and brown Naugahyde. When my guess proved to be the exact, predictable, scripted ending of the show, an angry member of the Class of '79 stood at the front of the small room and threw the bottom pillow of his club chair, Frisbee-style, at me.

"You bitch. You ruined the whole thing," he yelled.

I think he meant the TV show, but his anger measured more in terms of ruining the whole academy. I stood silently and looked at him in total confusion. I turned my head slowly to look at the other men, some classmates and some from the other two upper classes

C3Cs (Cadets Third Class) Kathleen Utley and Mark Kaspar heading to afternoon military drill (marching practice). Mark as Guidon Bearer carries the 17th Squadron flag. I'm with "white gloves, under arms" (courtesy Barry Staver Photography LLC).

that were sprawled on the seats. Only a few even turned to look at me. Lip bleeding from the pillow zipper across my mouth, I left without any word of support from any of the only male eyewitnesses there. There were still battles to be fought.

15

Two Degree

"When peoples care for you and cry for you, they can straighten out your soul."
—Langston Hughes

"Get up, you lousy Basics. Get up. Full fatigues and out in this hall in five minutes. We are going for a morning run to greet another BCT day."

Was it another nightmare? It was familiar in a terrible way. The sound of the fists on the door. The shakes and sweaty skin as the mind returned to reality from the dreams. I could remember the cold floor on my bare feet before my socks and tight black combat boots constricted them for the morning run. I felt it all still. I heard it too.

But it was no nightmare. Now a "Two Degree," with only one class senior to me at the academy, I was finished with the awful dreams of that first horrible BCT summer. These screaming words were coming out of my mouth. As an element sergeant for first BCT, June 1978, it was my turn to be the yeller. My turn to bang on the doors, and these Doolie women could not escape behind its wood like we could because they were not "decent" yet. I jumped right in and got close to their faces whether they were dressed or not.

"Get moving, ladies. Get moving. You are not going to make the whole element look bad by being the last ones ready," I bellowed.

It was their monstrous summer. It was a lark of a summer program for me. As cadre, I got up ahead of the Basics but with my own alarm, no one beating on my doors. I dressed in fatigues but only the heavy pants. I wore the white USAFA T-shirts of cadre members instead of the hot long-sleeved fatigue shirts. I had brushed my teeth, brushed my hair and had time to apply a quick bit of mascara if I felt like it. I stood poised outside the Basics' door chatting with other upper-class cadets waiting for the second hand to reach the exact minute for the jolting beginning of the tirades to start their day.

I knew this BCT that it was such a mental game. Of course, my legs

15. Two Degree

were not convinced of that when I was jumping out of my bed as a Basic. I ran the same distance now. More, actually. I ran back for stragglers, ahead to chat with classmates, back to urge on the Basics in the back of the formation and up to the middle again. It was easy. Yes, my lungs had acclimated to the more-than-mile-high thin air. My legs walked around the terrazzo daily, and I participated in intramurals. Most importantly, my mind had become accustomed to the challenge and had stretched to include extra possibilities. I knew what to expect. I knew how far we would run. I knew what the day held. I recognized what my body and mind could do. I knew who I was and that I belonged at USAFA.

The Basics had no idea anymore who they were. They were still being stripped of all the high school honors that made them someone special. They were not quarterbacks, valedictorians, National Honor Society members or sweethearts anymore. They were Basics, and it was my job to keep reminding them of that. Training brought them down to a fundamental level so that they could be remade into Air Force officers during the next four years.

Karen Wilhelm talks about some hard lessons learned at the academy that had to be unlearned. She "learned to avoid showing weakness or vulnerability," not a good thing, Karen said. For her, the result was hesitating to ask for support when she needed it. "Worse than that," she went on, "it meant I sometimes didn't offer help or support when others needed it."[1]

I think I tried to support the Basics. I don't think I treated the women more harshly or any easier than I treated the men. I wanted them all to be Basics. I knew I couldn't be seen to be easier on the women who were one and two years behind us. The concept of women at the academy was still being evaluated.

With all the new Basics, I pushed them and taught them, and I cared about them. I worked with them for the three weeks of first BCT in the cadet area and handed them over to the second BCT cadre for their time in dusty Jacks Valley. At the end of the summer, just before the school routine began again, I was proud of all of those left from my element on their day in August when they were accepted as Doolies and official members of the cadet wing. I also knew that they would not all make it to their Two Degree year.

My Two Degree year meant big things for me and for the entire Class of '80. This was the year we would get our class rings. This year, we would get our cars. The Class of '79 still ruled the terrazzo, now in their senior, Firstie year, but we were next in line. We moved back into our squadrons with the same '79ers, ourselves and the new Three Degrees and new Doolies holding on to the bottom rung. We no longer worked directly with the Doolies in the squadron. They were the responsibility of the third classmen. We moved into positions with more responsibility of truly running the squadron under the leadership of the Firsties. We had all come a long way since our Basic days. The women had come a long way in the two years since our training under the ATOs in the "penthouse."

My two best buddies held the squadron's highest Two Degree jobs. Bob Schley was the training sergeant, a job he well deserved. He was in charge of overseeing the weeknight training of the Doolies and the weekend inspections of our rooms and rifles and uniforms. He also had to arrange for drill practice so our marching looked uniform and synchronized when we marched in parades or even to lunch. Fred Kornahrens, as first sergeant, was the highest ranking cadet of our class in our squadron. He was the leader of our class and responsible for schedules for all the squadron duties, for the discipline of the lower three classes and to oversee all squadron activities basically as the assistant to the cadet squadron commander. His black felt shoulder boards were filled with silver chevrons, a very visible status. I was an element sergeant again, as I had just been during basic training for the new Doolies with minimal silver of rank on my shoulder boards. But that is the way the squadrons worked. Each year was divided into two go-rounds. First go-round, one group held the higher positions, and the next "go," another group would step in. Every cadet could have the opportunity to have some leadership role at one time during the year. I wish I could have served with Bob and Fred. They were fair and funny. They cared about me. We would have made a great team, but I would wait my turn.

In the meantime, academics kept us busy. I was a computer science major and spent many hours in the comp sci lab over in the academic building. We were still required to be in our rooms by 2300 taps, so I sent my last programs through the card reader by 2250. Then I ran with my books back to the dorms to say hello to the DI when they came by for their nightly headcount. There was no rule for the upperclassmen to be in bed or asleep, so the homework from my other classes waited until after 2300. We also all marched in squadron drill every other afternoon, alternating with intramural sports. Drill at this point was very automatic. Step off on the left foot, follow the beat of the drum or the cadet yelling cadence. With or without rifles, I was comfortable marching. Bob was also the captain of the squadron intramural soccer team and let me be on the team. It was supposed to be a choice, but the team captains for the squadron built teams that would likely succeed through an unofficial draft. One year, we laughed that the 17th Squadron swim team were the ones wearing the life jackets. The strongest athletes were on other teams to maximize winning. I technically should have been with the float-assisted swimmers, but Bob knew how little I liked the water and had pity on me. I didn't do a horrible

15. Two Degree

impression of a soccer fullback. My knees took a few good blows, and I spent some time on crutches.

In the roommate department, I wasn't going to cruise this year. Louise had left. The change to the Armed Forces Veterinary Corps ended her plan to be an Air Force veterinarian. She also fell in love. She got married in May at the end of our Three Degree year and asked me to be her maid of honor. I had made my own dress, wore a crazy hippyish floppy hat and popped champagne in her new backyard last May. Her groom was her academic adviser and biological sciences instructor. More than a few windows rattled with the rumors and then confirmation that this female cadet was marrying an officer. Her husband told me that they were only discovered because one of the department administrative assistants happened to see a notice with their names in the local paper in the marriage license column.

But in her serious, unshakable way, Louise seemed sure and marched on through finals and accepted the low quarters shoes and fatigues, the uniform of the "quitters." She walked down the "Bring Me Men…" ramp with her head high. She took all her academic credits too, with no requirement to pay back any of the education with years of service or money. That was the deal with the academy. Up until the end of your sophomore year, technically until you declared your major on the first day of academics as a Two Degree or junior, you could leave without any obligation to join the Air Force. With all the 20-hour semesters we carried, Louise enrolled in the University of Colorado, Colorado Springs as a second-semester junior and would graduate and go on to a college of veterinary medicine before I would finish at USAFA and start my Air Force career.

And she seemed happy. I knew. I still saw her. He still taught, and they lived just downtown outside of the south gate of USAFA. We got together on some weekends. I got a chance to sit in their kitchen and help to cook a meal or hang out and watch a movie with them. I lost a roommate but now had a civilian friend downtown with wheels. It could be fun having a real civilian girlfriend.

The time flew by. The fall semester ended with finals and everyone catching buses to the airports for trips home. After the new year arrived in 1979, it was still hard for me to return but not the heart-wrenching hard that it was my Doolie year. It wasn't even the exhaustion-filled return of the Three Degree year. I was coming back to friends, to my computer science and math classes, to comfort even in the uncomfortable. No one enjoyed Saturday morning inspections, marching on the terrazzo in the bitter cold, running through the fitness tests where the trash cans awaited you at the end in case you couldn't hold down your breakfast. It was still grueling, but it was mine. My academy, my education, my squadron and group and, sooner than I thought, my real Air Force. I came back almost excitedly.

I was rewarded for all my military efforts with the small silver insignia of the Commandant's List, recognizing leadership, to wear on my uniform. I was also offered a position on the group staff, the cadre of cadets to lead one-fourth of the entire cadet wing. I hated leaving the comfort of my buddies of 17th Squadron as I would have to move to a room downstairs in the group staff area. USAFA, I knew by then, is not about encouraging comfort, though. It is about challenge and achievement and competition. I would be living apart from my squadron mates but in a position of authority. I earned the slot. Based on my military standing, I deserved this job and took it without hearing any whispers about tokenism, if there were any. I packed up and moved down two floors and over one quadrangle to join the selected male cadets there to run the second group.

On group staff, my schedule changed a bit. My responsibilities were different, but the purpose was the same. As a group staff, we managed the ten squadrons in our group. For second group that included squadrons 11–20. We oversaw more people but still to the end of developing fine officers for the Air Force and upholding standards and discipline. I marched to meals behind the cadet group commander in a group of five, not 100, and felt the presence of the ten squadrons of the group following behind. I inspected rooms and held meetings, learning more about command than teaching about following. I still got back to the squadron to go out for ice cream with Bob or Fred. I still went to 17th Squadron parties or to Louise's for some R and R. On my birthday, though, it was the cadets of group staff that pushed me out into the snow in the quadrangle and locked the doors, laughing as they ran away. I laughed too. It would have hurt, more than the cold snow on my bare feet, if no one had played a prank on me. I was an Air Force Academy cadet, and I had acclimated to love its traditions and camaraderie. At the same time, I lost touch with my closest friends and with any of the women. I could have used their guidance and support when I got caught in a scary place alone.

I knew a man, not a cadet, in a very official capacity who worked at the academy in an Air Force uniform. He was friendly when I saw him. He was divorced, and he talked often about his children. His asking me to come to his off-base house and babysit did not surprise me. I was responsible and friendly, and I always showed an interest in his stories about his children. I loved getting away. I thought it would be like heading over to Louise's relaxing for an evening, so I agreed.

He picked me up from the cadet area on a Saturday evening and drove me to his home in Colorado Springs, not far from the academy's south gate. I got the meal and bedtime instructions for his children and settled

15. Two Degree

in for a babysitting experience to add to my academy history. The young girl and boy cooperated, minus the expected fuss at bedtime. I relaxed to watch some TV, in my jeans, away from USAFA, like a real young woman.

He came home happy and friendly and probably a little drunk. He offered me a black cow. I could only picture some kind of chewy candy that I ate as a child.

"No," he said, "it's a drink."

"I really don't drink," I explained, almost embarrassed that I wasn't that sophisticated.

"But it tastes like a chocolate shake," was his reply.

With a drink for him and the "shake" for me, we sat on the plaid couch to talk a bit. I tried the cow cocktail. The aftertaste of the alcohol burned my throat. Trying to hold the liquid out of my mouth, I fished a piece of ice from the glass. I hid the drink away under the arm of the couch.

The conversation got too personal. I didn't want to discuss my hair or my eyes. Discomfort rose from my heart. My stomach flipped once, twice, more in warning. The burning, not from the drink, boiled up through my esophagus and clogged the back of my throat. Swallowing didn't relieve the lump there. My hands felt cold, and my legs jiggled with nerves. I stood.

"I need to go now," I said before my knees were even straight.

"I thought you said you were free for the evening when you said you'd babysit," he answered.

"I was. I am. But I don't feel very good now. I just need a ride back."

Standing, what I thought was defiantly in the middle of the room, I waited. He rose and came toward me. He reached his arms around my small frame and pulled my body toward him. My reactions were slow. All the unarmed combat training and assault course bayonet stabbing had not prepared me. My mind was soft. The situation seemed to register unbelievably slowly. I pushed back and looked up to see what this all meant. His head moved in front of my eyes, and he moved his mouth toward mine.

Adrenaline finally reached my arms. I would not be a placid doll. I stiffened and pushed back enough to break the hold.

"I have to go," I whispered. "NOW!" came out louder.

I wanted to run or scream, but some kind of military mindset kept me in a regulated state. In the presence of a superior, I felt required to be obedient, maintain decorum and follow orders. I got my coat, headed for the door but stood and waited for him to recover and come toward me. And nothing else happened.

He drove me to Louise's where they were expecting me to stay the night. Cuddled up watching a movie, Louise and Bill invited me in. I headed straight to their guest room feigning a headache. The potential danger of the situation fell upon me as the guest room door closed. I collapsed on the bedspread, pulling it around me like a cocoon. And I cried. And I shook for fear of what might have happened. I yelled at myself for being such an inexperienced fool, for not seeing this coming. Until morning, I worried about what would become of my career and how I would be able to see his name tag on that uniform and act as if nothing had happened. I wondered how I could do anything but ignore the whole thing. I was ashamed of my own naivete.

The next morning, my mind was still muddled. Louise drove me back to the academy so I could attend Mass and catch up on homework. My thoughts turned to the tightness of the hug and the closeness of the mouth that I didn't want. I couldn't think.

More scared of Monday morning than if it held a ten-mile uphill run with rifles, I struggled to figure out how to go on without ignoring the whole episode. How should I react? How should a military subordinate react to a superior's unwanted advances? There were no quotes in *Contrails* memorized in my head to guide my decision. There were no indexes in the regulation book to reference. There were no upper-class women cadets to ask and I didn't feel like I could turn to the female AOCs for guidance. I could not yet feel the strong bonds with the sisterhood of my own female classmates to call on for strength and direction in a situation like this. We had been trained in unarmed combat in Jacks Valley, trained in 15 count rifle manual arms, but I cannot remember any training about how to handle unwanted advances, situations of discomfort within the imbalance of rank in non-saluting, non-marching circumstances.

I decided to call an officer who was an Air Force lawyer. I didn't know him very well, but he was part of the academy "family." We had home numbers for many officers, instructors, AOCs and liaison officers who were extra mentors for our squadrons. I cannot remember why we had all their contact numbers. I know we could call our instructors to schedule extra instruction for classes, but it seems that would have been their office numbers. I know we could call our AOCs at home if an emergency came up. We could call without repercussions. My call was not to a part of my chain of command. That seemed too official. I wanted some control, some discretion in deciding my actions, and this officer whom I already knew as kind, caring and objective was my choice for help.

I kept my composure enough to get through his wife's answering the phone but broke into tears telling the story and giving words to my fears. He came immediately and picked me up. Looking back, I don't think I told him about the offered drink. It was a drink I didn't ask for, and I had only a sip but was still underage except for 3.2 beer. Underage drinking was a punishable offense for cadets. I was honest about my dopiness and lack

15. Two Degree

of understanding about the circumstances. Telling the story brought back my fears of the moment, of the assault that could have been. I was afraid of repercussions from pushing that other uniformed man away. I was afraid for my reputation if the story got out. I had worked hard to be a disciplined cadet, earning my rank. I could not stand for a reputation that I bought anything with any special favors.

He advised me in a calm, reassuring voice. I would be his "client" with attorney-client privilege, even though there was not a formal case. That would come from more official procedures and exposure that I unambiguously did not want.

I trusted him. I wanted to turn the whole situation over to someone else. I wanted to not have to think about it anymore. I wanted it to go away.

The lawyer went to see the other man to gauge his position on the incident. He stepped into the man's office, explaining that he was there on my behalf. He was wearing his formal Class A suit coat jacket with scales of justice in the JAG insignia added to the equation. The man broke apart and admitted he was wrong, and he apologized. He apologized many times.

The JAG officer relayed the situation to me. If I were comfortable with the assurances that it was an isolated mistake, that the man was truly apologetic, that he would never forget his decorum again, we would put an end to the whole incident. He did not pressure me. If I wanted a formal inquiry, I could have that too. It was my right.

I definitely did not want a formal inquiry. I was tired of the challenges. I was afraid of what would be said about me. I felt any scandal would tarnish all the women at USAFA if the story came out. I couldn't stand to think anyone would judge us all to be flirty or trashy if they heard. I couldn't stand to have more derogatory words added to the things I was called. I knew there were men waiting to hear of our failures and would blame any bad publicity and bad outcomes on the women of '80. I wanted to put it all away and move on.

A man in a uniform had crossed the line. He scared me. I lost my feeling of security because he lost sight of his responsibilities with his position of authority. He set back the growth of my confidence that the trials of the academy had built up in me. I could blow the whistle and mar his career. I could punish him for forcing himself into my space and my life.

Another man, in a different uniform, had treated me with respect and dignity. He listened and cared. He acted in my interest, as my advocate. I trusted him. With total professionalism and compassion, he represented and advised me.

I let the goodness of one outweigh the inappropriateness of the other.

I received a verbal and, seemingly, a very sincere recognition of guilt. I satisfied a 20-year-old's need for closure within the boundaries I understood at the time. But I can now trace a line back to some responsibility on me to forge a safe environment for the women who would follow.

I continued to interact professionally with both men. One, I saw, but I never talked to again, beyond a formal military exchange. The other stayed in my mind as the perfect kind of Air Force officer, the kind that I hoped that USAFA was shaping in me.

16

Señorita

> Tengo un día. Si lo sé aprovechar, tengo un Tesoro.
> —Gabriela Mistral

There were exciting programs at USAFA that introduced women cadets to the community, the Air Force and the world.

Women Air Service Pilots (WASPs) of fame from their flying service during World War II had a convention in Colorado Springs. As a cadet, Kathy Conley and some other female cadets went to lunch with the WASPs at the famous Broadmoor Hotel. The famous flyers became aviation role models for Kathy. When upperclassmen spewed the threats that women would never be pilots, they wouldn't be able to learn to fly, Kathy had her retort. Probably not a verbal rejoinder, but she had an example to hold in her mind of successful female pilots to add to her aspirations.

Kathy also participated in an exchange program with the French air force academy, l'École de l'air, in Salon-de-Provence. After intense language studies at USAFA, Kathy Conley, another female cadet (Debbie LaFrombois) and six male cadets from USAFA went to spend a semester with their French counterparts. The French cadets were very welcoming. The American cadets were invited to Monaco to meet Princess Grace who was curious about the American female cadets and asked them if they intended to fly. She referenced Amelia Earhart as a female pilot she met as a girl. The cadets marched down the Champs-Élysées in the equivalent of a Veterans Day parade. Kathy even had a French spirit mission prank pulled on her and woke up with her bed completely flipped on top of her.

The école had women in the ranks. Three French women bunked with Kathy and Debbie, but their training wasn't the same. The French female cadets were excluded from some aspects of the program because France did not allow the French women to become pilots. Kathy came back to USAFA with flight experience and a certificate of completion from their parachuting program that her French female colleagues could not get.

In the summer of 1979, as a rising Firstie, I was out of uniform. In a lacy peasant blouse and a skirt, nearing midnight, I was headed to a discotheque for some late-into-the-night dancing. In the lights and bustle of Buenos Aires, I walked in heels with a few Argentine civilian young women and about 20 male cadets. The ratios were perfect for plenty of dance partners and plenty of salsa and disco moves. Half the men were fellow USAFA cadets and half were cadets of the Escuela de la Fuerza Aérea Argentina, a South American sister academy.

I too participated in an exchange program, shorter but similar to the French exchange, through USAFA with the Argentine air force academy. Ten male cadets and I gave up our three-week summer break to go to Argentina. I was the only woman interested or the only one with enough Spanish-language ability or the only one crazy enough to forgo time at home that summer. The Argentine academy, located in Buenos Aires, had no female cadets. We left USAFA, Colorado Springs and headed to South America on an Air Force C-141. We traveled in uniform, but my packed suitcase held some jeans and some civilian clothes for some time off in Buenos Aires, Mendoza and Bariloche.

It was a glamorous, glorious time for me with Latino cadets holding doors, carrying my suitcase and opening car doors. I danced more hours than I slept. I was the belle of the Buenos Aires ball as the American cadets seemed to watch wondering what the fuss was about. They were interested in meeting the Argentine civilian women, but not many were available during our military tours. I reveled in the attention and gloated about how the American cadets did not realize what they had. They were all my buddies, but not a one ever asked me to dance back in Colorado Springs. "I Am Woman" from Helen Reddy came flowing back into my consciousness, minus the demanding beat. It became a romantic song. I danced through the three weeks in Argentina dreaming that my remaining year at the U.S. Air Force Academy could be so idyllic.

What I didn't realize was the trade-off was absolute. These Latin men treated me and accepted me as a woman and a female companion. They all, Pedro and Luis and *los cadetes,* treated me with typical Latin male *consideración* and *cariño.* They did not ever ask me any air power questions. Unlike Kathy and Debbie, I did not get flying or parachuting opportunities. The cadets never invited me to collaborate on their engineering homework or with any class project. They kept me out late at the discos and bid me goodnight saying there was no training for me in the morning. Then they left me to sleep in while they took their morning runs with my male classmates, Tom, Ken and Mark. They brought me flowers but did not offer me a place on the *fútbol* team for practice. I was so caught up in the romance of the attention that I overlooked the lesson of the missing equality.

16. Señorita

USAFA Cadets and Cadets of the Argentine Air Force Academy, July 1979 (author's personal collection).

My picture was in the Argentine newspaper with a headline of "A Woman in the Air," even though I wasn't pilot qualified. My comments to the reporter when asked about what I thought of the lack of Argentine women in their academy was an acknowledgment of the cultural differences in our countries. "The Argentine woman is in evolution," I said, "and will join in the future in tasks that are currently closed to her. Many of the activities that the female sex carries out in North America but not yet available to the women here, does not mean that Argentina is backward, but rather that there are different customs of the peoples of Latin America." I had lived for a few years in Costa Rica. I knew women's roles were different. I experienced some of the cultural differences, and even as a teenager, I liked some, but others felt obsolete. I didn't share my thoughts. I think I was briefed on what to say.

When I got back to USAFA for my final year before graduation, my classmates, male and female, held doors for me. I held doors for any coming in or out behind me. It wasn't gallantry. It was proper academy behavior and common courtesy. On my return to USAFA, I accepted drill over dancing.

The Argentine air force academy opened to women in 2001.[1]

17

Firstie

"By three methods we may learn wisdom: First, by reflection, which is noblest; second, by imitation, which is easiest; and third by experience, which is the bitterest."
—Confucius

Our Firstie year was supposed to be the best. I had a brand-new blue Toyota Celica in the parking lot with my name on the title. Fort Sam Houston Bank actually owned it, but I had the keys and the payments on the loan were deferred until after graduation. I was more practical and conservative than most Firsties. Groups of cadets haggled directly with some of the nation's largest car dealerships and manufacturers in the Detroit area. Cadet clubs, formed around the cars they were buying, met beginning the previous winter to plan the purchase of large lots of each type of car. "The Class of '80 Firebird club will meet at 1900 hours today in the Twenty Third Squadron Assembly Room." "Any cadets interested in buying a Camaro meet Wednesday at twenty hundred hours in Arnold Hall." Dealers dealt. Cadets flew off to Detroit to pick up 250 Firebirds or 200 Camaros. They convoyed back to USAFA with their new wheels and new freedom. Fred bought a beautiful white 1979 Corvette with blue leather interior. Bob got a red Corvette. John Ward bought a blue one, and the three parked side by side in the cadet lot looked very patriotic.

With the Class of '79 gone, we felt we owned the place. That particular group of cadets had held on to their special place in the history of the academy as the last class without women to the bitter end. Some even had the notorious LCWB, ostensibly for Loyalty, Courage, Wisdom and Brawn, inscribed clandestinely on the insides of their rings. The letters would have made it to the crest on the outside of their class rings when they were Two Degrees, but the rumor mill leaked an alternate meaning of the acronym. The letters were categorically banned. LCWB: "Last Class with Balls." It was not an appropriate or approved class motto. The letters went underground, but the sentiments about their being the last vestige of manhood, and therefore worth, at the academy did not completely submerge. I'm sure the men of our class were not happy with the implication of manhood leaving USAFA with '79. By then, I didn't like the implication either since I had grown the "get in your face, don't mess with me" kind of balls. But the Class of '79 produced some stellar cadets. I liked many of them. I knew they would make wonderful Air Force officers. However, I was not sad to see them toss their hats at graduation and roar off into the sunsets in their one-year-old cars.

Command belonged to us now. The Class of 1980 finally made it to the top of the ladder of USAFA. We were Firsties with all the ensuing privileges and responsibilities. Despite others' doubts about our "Burger King—have it your way class," with all our "weak" positive motivation for training, we were ready.

I was ready. My body and spirit felt prepared. I was finishing up the requirements for my computer science degree. I was in Argentina last summer. Then I had worked with a great fun team of cadets to run a camp for underprivileged children at Homestead Air Force Base in Florida. Even without any leave time at home, I was set to face this final year.

I had a secure place in 17th Squadron. Bob and Fred were roommates and lived next door to me. Ron and John were good friends too. Ron, Bob, John and Fred included me in part of their circle of compatriots.

There were women in all four classes. I wonder how it felt to the Class of '83. The academy they joined had women in every class. There would be no other reference for them, as if it had always been that way. How odd.

The Class of '82 was in charge of training the new Doolies. I did ask some knowledge questions in ranks. "Mister" or "Miss, when's the next Air Force Day?" was my favorite. It kept Doolies on their toes about dates and events in Air Force history, but also it was forward thinking. The next thing in our calendars. Not history or aircraft but moving, thinking and memorizing forward. Closer to graduation for me. Closer to recognition and, in four more years of Air Force Days, closer to graduation for the Doolies.

I cruised with the Firsties of Seventeen. We walked together to meals and sat at the heads of the Mitchell Hall tables. We talked about post-graduation assignments, pilot training for many of the men and women but not me. Unless the Air Force authorized high-heeled boots, I was too short to fly. If I pulled the cockpit seat up close enough to reach the rudders, my knees and below would be left behind if ever I ejected. When military schedules allowed, we Firsties cruised downtown too. We went out for pizza or to a movie. We attended squadron parties and had plenty of privileges to stay out or come back late. But we still studied and always attended class. It was an offense punishable with demerits if we skipped.

In January 1980, I returned to Colorado Springs for one last semester. Right away, I was required to report to the officer in charge of the second group for a meeting. Having served on group staff last year, I had my suspicions that he would ask me to serve again. But I also expected a position of leadership in my home squadron, Stalag Seventeen. I would take most jobs in Seventeen over a secondary staff position on group staff.

"Come in, Utley," was his answer to my knock on the opened door's jamb. "I'll get right to it. We are considering cadets for positions on the group staff. I am impressed with your record both academic and military."

"Yes, sir," I answered, proud of his recognition but fearing the worst. I did not want to be group logistics officer.

"Utley, I want you to be commander of the entire second group this semester."

His words caught me by surprise. I expected him to ask me to be on group staff. I thought I had earned that consideration. My grades were high, my military standing at the top of my squadron, my record clean.

But group commander was beyond my expectations. Group commander was the head cadet honcho for ten squadrons. The group commander was in charge of more than 1,000 cadets, one-fourth of the cadet wing.

I shifted in my seat. I pulled my spine up straighter and off the vertical back of the lieutenant colonel's office chair. My mind started playing the pictures of me, marching to lunch, in my solitary row, flanked by my staff with the entire group of ten squadrons in step behind. As group commander, I could be the first and only, not just the first. Not one of 157 other women but the first. I would stand apart. I would make bigger leadership decisions, be included in meetings higher up the chain of command, separate myself from the rest. I imagined my father's pride. I saw my cadet colonel shoulder boards so full of silver that none of the black felt showed through.

And then I thought about the battle, the fight of these last four years to earn a place here. Not just for women but for me. I thought about other male cadets that I knew who had fought for their place here too. I knew I would have to walk back into the squadron with my news and be able to face the male cadets who had become my friends and my sounding board. I could be group commander, but I couldn't sell it as fair to the men I most respected.

I hesitated, semi-convinced that I knew what to do but not solid in any decision. I know my face betrayed my indecision, but I don't remember waiting too long to answer. I know I would not have kept a colonel waiting for a response. What I thought I should do and what I thought I could do bounced between the walls of my skull and confused me more. The shock of the offer must have been the deciding factor. I would not have been so stunned if I had thought I truly deserved the consideration. In so few seconds, I knew I could not take the job.

"I want to thank you very much, sir, for the consideration. And although I am very honored to have been asked, I just can't. I know of men in the group that are more qualified." *I cannot be a token. It will erase all that I have fought for.*

"It's time for a female commander, Utley."

"Maybe, sir. It just can't be me," I responded. I thought, *He has his own vision of what will solidify the women here. I have another.* I could not in good conscious take this position. "I hope there are no hard feelings, sir," I said out loud.

Dismissed, I saluted, turned sharply on my heels and left the officer with my decision.

In retrospect, I wonder why he didn't counsel me more. Why wasn't there consideration from his side of the desk about my taking some time to think about it. I didn't use the word "token." He didn't use it either, and thinking about it with age on my side, I realized that in the all-male, mostly white ranks of the cadets and the officers, the idea of tokenism and the ramifications had not germinated. No experience of it or thoughts about it might have ever entered their minds.

Back in Seventeen, Fred and Bob thought I was crazy. "He will just pick another." "You should have taken it. It happens all the time." "Lots of male cadets get jobs when they are not the most qualified." "You could have done it. You could have done a great job and proven even more." "You could have been the group commander and gotten us all extra privileges and released us from every inspection till the end of the year." "You could have gotten extra parking spaces for us closer to the dorm." And then their words were nothing but kidding.

If I had waited for their support, I might have decided to accept the command. But what kind of commander would I be if my confidence came from only the support of others? And theirs was not indignant overwhelming support. It was questionable support. Kidding masked some of their insecurity with their convictions. If I needed assurance from myself and still from others to stand strong as the commander so visible to all, the topic was closed.

My air officer commanding, Henry "Hank" Wilborn, told me years later that he knew I was going to be asked and he was trying to save me from being a showpiece. He offered the facts to the higher-ups of another cadet, that a male cadet had higher military standing than I.[1] As an African American pilot, he had faced doubts himself from others. Not whispers but direct accusations of tokenism, questioning whether a Black man could be a pilot without affirmative action to earn a slot. Standing at the Coke machine at his Undergraduate Pilot Training (UPT) break room one day, he overheard other UPT classmates talking about who they predicted was going to "wash out," failing the program. One of his fellow officers and pilot trainees said, "The Blacks aren't going to make it. They don't belong here in the first place." Unfortunately, he wasn't surprised by the attitude. He had heard it all before. It did further embed in him a strong resolve to succeed.[2]

Major Wilborn proved himself worthy in a C-141 and a B-52 of flying, of commanding and of being an Air Force leader. He was translating his fight to be accepted on merit and not as a token to our fight, to my fight to be accepted as a cadet and not a female cadet, not a token.

The officer did eventually choose another woman. She might have been the most qualified in the group for the position. I knew her and thought her sharp in her military and academic accomplishments. She did a good job. I felt some stabs of envy in my throat, though, when she called the group to attention during parades. My pulse raced a bit when she sat high on the staff tower, but I cringed when I still heard rumblings about tokenism.

I had yet to learn that tokens can fill voids, change perceptions, hold places and be role models. Gail reminded me that "what you can see, you can be." Female cadets could have seen me as their model for command.

I made my choice based on what I thought would be the men's perception of me as a token. I did not want any charity and gifts for being a woman. I wanted what I thought I had earned, no more, no less, and wanted to be lauded for that level-headed thinking. What I did not realize until years later was that even if it is a gift, the impact is not always just the gift; it is what you do with it. The officer may have been giving me something, but it was not a stretch. I could have done an excellent job. I could have been the one to share that gift of women's leadership with the women in the underclasses. What you can see, you can be. I could have shared the gifts of the lessons learned from my leadership mistakes with the academy and its future. I was so busy being one of the men that I forgot to be who I was and to grow to be who I could be. But my doubts were maybe reason enough for the colonel not to press me and another woman did a fine job and continued our work to open the path within the men's environment. I can now say with the years of experience behind me that I could have. Should have? In my truth of the moment, I am still not sure of the should.

My commanding officer asked me instead to be commander of 17th Squadron. That, I accepted without question. Fred was the only one who might also deserve it, but his grades had suffered after a Monday night football game ran into overtime and the engineering instructor didn't ask one single football question on the exam. Academically, I beat him out. He knew it was fair when I was chosen. I wore shoulder boards with one fewer bar than I would have on group staff. In formation, I still stood alone in a row but with only 17th Squadron behind me. I commanded one-tenth of the cadets but was 90 percent more assured of my authority and myself. Some still grumbled about tokenism, but I felt steady on the foundation of my qualifications.

But authority and responsibility did not come wrapped in a neat defined package for me to open and administer. With all the rules and regulations in the giant notebooks at the academy, I faced one situation not as black-and-white as the typed words of the regulations. I struggled to find the solution on my own.

A cadet came into my room on a weekend morning. "What's up?" I asked, the look in his face betraying his turmoil. "I want you to see something." He moved to the window and a familiar blue Corvette drove up the north road and turned left toward the parking lot beneath our view. "That's John," he announced. "So?" I questioned. I knew John's car, and we could see John's face framed in the driver's window before the turn. "Watch the rest," the cadet demanded. I now felt anxious about the watching and about this cadet's temperament. He was usually such an easygoing guy, a well-respected cadet but not overly strict. He got the job done and still knew how to have a good time.

But he had come to me a few weeks before with a dilemma. He suspected a few of the Doolie women of dating upperclassmen in the squadron. It wasn't the most heinous crime in my book. I had done it myself. But this cadet had a right to enforce regulations, and I had a responsibility to him and the squadron. I called in the accused parties in separate groups and did my best to smooth the situation away.

The two Doolie girls were the easiest to counsel. "You cannot do this anymore," I told them. They stood at attention in my room. I sat at my desk chair. My authority over them was easily recognized in my rank, their lack of rank and my comfort. "Your fraternizing is leaking out into the open. I cannot ignore it if I see it," I told them. I said I understood their need for something kind and caring in their Doolie year environment. I think one of them even repeated what someone had told them: "You did it, Cadet Utley." I would rather not have had that in the conversation, but I did not deny it. "Yes, I did it, and it was against the regs then too. I didn't get caught, though, and that will be the difference. I am not on a witch hunt here, but rumors are flying around the squadron on this one. I won't be able to ignore them forever. I will not cover for you."

Subdued and serious, they reported out of my room with their salutes. I hoped it was over.

I spoke to the two male cadets too. Not together. One was a friend of mine and I thought deserved more of my personal consideration. The other, a Two Degree, got little sympathy from me.

"You are messing around here, and you are going to get burned. It has come to my attention, and I am generously offering you this warning. Don't do it. Don't let me see it, or hear about it, or know that it is going on if you continue. I will have no choice but to write you up."

With John, though, I restrained my tone. I felt more cautious around him. John was part of my group of buddies. He was not the closest to me. He was good friends with Bob, Fred and Ron. I appreciated their friendship and support. Sometimes they just teased me, and there were times I could not fit in with their camaraderie because of my gender, my interests or my experience, but I considered them friends. I treasured their acceptance and company.

John also was in one of the most awkward and unique situations of any cadet I knew. He wore consequential residue of his years in the Class of '79. John had started at the academy a year before Fred, Ron and me in the summer of 1975, a Basic cadet in the Class of '79. There were no women in his BCT or during his whole first Doolie year. Well, that is not completely true. During John's first year at USAFA, the ATOs arrived. John saw them, officers, imitating Doolies while he was a real one with all the strife that accompanies that lowly classification. He was a classmate who saw the very first women march into the cadet wing.

John had a very defined and very intense sense of determination at the academy. He was the son of a military man who served time in Vietnam flying helicopters. John had lived with a parent in uniform or in the absent shadow of a father who was fighting in the jungle long enough to have the desire to serve became his destiny. The only son of the family, he had lists of duties to accomplish while his father was deployed. At age five, he was so awed by "the song that could stop cars," our national anthem that obliges cars on military bases to stop for taps. He told me it upset him at a very young age if a car failed to stop, or the occupant did not immediately get out of the car to face the flag and render the required salute. He had it in his life plan by fifth grade to go to West Point, join the military, "fly jets and kill commies." At age 16, with his name on a private pilot's license, his dream morphed from West Point to her sister academy, USAFA.[3]

As a Three Degree in 17th Squadron, John trained the women of '80 there. He admits that there were women who were very sharp. Terry Armbruster impressed him with her sharp salutes, her accurate knowledge from *Contrails* and her athletic abilities that added squadron points. He encountered other female cadets who did not impress him at all. I know I felt the same dichotomy about the male cadets I encountered.

17. Firstie

When I arrived in Seventeen the following year, however, John was no longer a cadet. During the summer, John participated in the airborne program at Fort Benning, Georgia. He was stung again by the differences because of the presence of women. "The men had to shave their heads as an airborne tradition," John said, "but also with an explanation that the helmets were more secure and better fitting without the hair beneath." The women just tucked their hair into the helmets. He qualified for the program by being able to do 12 pull-ups. The women's requirement was a flexed arm hang that seemed tepid in comparison and definitely unequal. Not during a jump out of an airplane but diving into a pool during his leisure time, he injured his neck. He did not return to USAFA in the fall.

Through recovery and rehab, he never gave up hope of returning to USAFA. He went to Rutgers for a year and fought on. He studied hard and wrote seemingly millions of letters to get his records reviewed for a return to the cadet wing. It rarely happened, but he wanted it. He won the fight, yet he came back to 17th Squadron, not with his buddies of 1979, but with the Class of '80. From the last class without women, he became a member of the first class with women.

Now John was dating a female Doolie, and I was supposed to straighten out the offense. It was problematic. He still wafted the aroma of '79, the class so proud to be all male. He was hardened and tough. John knew of real-life tragedies, of injury and recovery, of a fight for USAFA. He seemed a more dedicated, more determined cadet than any, male or female. He intimidated me and inspired me.

I told him that rumors of his fraternization reached my ears in my official capacity as squadron commander. I gave him his warning in an official voice. I told him as a friend that I would not look for him and her but that they should be careful, for others were looking.

How would the fraternizers react? If someone had confronted me in my Doolie year, would I have given up the security of knowing someone cared about me every day that year? Would I have walked away from the notes, the calls, the clandestine meetings and the involvement because of threats? I don't know. I was serious about making a place for the women of USAFA. Yet as a Doolie, I had been very timid about any ruckus. Would fear of discovery have scared me enough to give up the phone booth calls and the escapes from USAFA? As a Firstie, would I have listened to a classmate and semi-friend who warned me away from the comfort of an involvement? Would I have believed that a classmate would actually turn me in for an offense that she had committed?

I expected they would give up the fraternization. Not just for their sakes or for the sake of the academy. I wanted it to all go away for my sake. It would give me no happiness to write them up and hand down the punishments. I didn't want to look like a traitor to my friend and classmate. I wanted him to be in a happy, caring relationship. I wasn't searching for or craving another fight.

Nevertheless, on that morning, the witness and I stood at the window of my dorm room looking out at the roof of John's new blue Stingray. Our woman, Stalag Doolie, emerged from under the stairwell. She opened the passenger door and got right into John's car. It couldn't have been more obvious. I saw it all, and someone saw me see it. It made me nauseous.

I wanted command, but I didn't want this. I wanted the authority over the cadets in the squadron but not this responsibility for their lives. I wanted friends for support and relief from the rigid discipline. I did not want to enforce the discipline on them.

"You have to write them up," he said, yet I knew this other cadet had the authority to handle the situation. It took me years to realize how to delegate and empower people to take responsibility. I could have asked why he didn't write up the punishment form, but I felt nervous about looking weak by passing the buck. Throughout our years at USAFA, the women had lived under the microscope. So many actions of ours seemed to be under scrutiny.

Why did he have to pick her up right under the windows? In all my fraternizing rendezvous, I went through enough switchbacks and spy-like moves to at least think I wasn't being obvious. Was he testing me? Our friendship? My resolve? Blinded by love? It didn't matter. The fraternizing deed was done and recorded. I felt I had no choice, and that restriction irritated my mind and my temper.

Did I talk to John first and then my AOC, Major Wilborn? I can't remember. Either way, I thought the die was cast. Later, I realized I could use command to think beyond the rulebook. I should have asked for advice. I was worried about seeming weak. I didn't want to show any softness that would be attributed to femininity or deficiency. I wrote the disciplinary form 0-10 in triplicate and turned it in. The regulations stipulated the punishments. There was no leeway for lenience.

And from that day for many, many more through the wall of my room, I heard the ranting of some of my male classmates. "She is a shit. A hypocritical bitch." The same words now coming from classmates. "She did it and got away with it, and now she is punishing for the same offense that she escaped from." I couldn't even pretend that they were talking about someone else. There were too few women to choose from. And through the wall separating the rooms, I heard voices call me names. And Utley became F-ing Uterus. And Kathy became a name much worse. And friend became enemy.

I had a radio in my room. Turned up with its one black dial, I could

17. Firstie

have drowned them out with songs from America or Bread. The regulation metal headboard of my bed was against their wall, but unlike the Doolie days of required room arrangements, Firsties were living in a gentler era of room arrangements at USAFA. I could have placed my bed farther from their noise. I don't know why I didn't. I allowed their voices to chastise me. I never banged on the wall or went to the library to study. I didn't fight back at all against their anger. I was punishing myself for not being a good friend. Why was leadership challenged by friendship?

I walked to breakfast by myself. An underground power convinced the third-class cadet in charge of dining table assignments to arrange that some people never sat with me. I was ostracized from some of this group of brothers to eat and walk and stand alone. *Be careful what you ask for, woman.* Command can be a solitary job. I felt more alone than I had in the tiny black box of the SERE prisoner of war camp. I heard the demeaning yelling then, but it was training, for my own good. Through the walls of my room now, I heard animosity. It was training too, I guess. But it didn't come from some simulated enemy. It came from the hearts of those that I knew.

Weeks passed. One morning as I stood in the hall locking my door, Fred walked up and asked if I was going to breakfast. "Hi. Good morning. You know, you could have handled that better," he said. But he walked with me to Mitchell Hall and sat at my table. "You could have come up with your own response and not let some other cadet force you into an action." But he crossed the threshold of my door to talk to me. "Maybe you could have tried to explain the whole thing to the other cadet and given him a way to buy into something less severe." But Fred was a friend again. He would have made a great squadron commander. He did make a great officer and leader.

Eventually, we all rewelded into a new relationship. Fred drew us all together again, somewhat. The scar was there, though. Our group friendship was never quite the musketeer-like bond that we had before. Yet, John is now a great friend. I asked him when the memories of the whole incident had softened, a bit, why our friendship ever rebonded. He told me of his defiant conversation with Major Wilborn, our AOC. After the punishment, John refused to do his duties as squadron academic officer. He explained that all the "tours," the one-hour blocks of time marching in a square pattern in dress blues with the M-1 rifle held at one's shoulder, would rob him of the time to do his squadron job. "You *will* do you duties," Major Wilborn explained. He told John that he was responsible for his own actions and the resulting punishments were his to manage along with his other jobs. Tours were exacting punishments because they robbed cadets of what they cherished so greatly—time.

Major Wilborn also advised John to put down his anger. John certainly had reason for anger in his life. To his credit, he did as the major advised and buried the heated resentment and moved forward. His life has shown his integrity, his resilience and his resolve. He made the Air Force a career and made the career a success. His life and his family reflect his integrity. His friendship reflects his understanding of wisdom and of USAFA as a learning laboratory that taught us all life and leadership lessons.[4]

Many years later, John's roommate and my buddy Ron McNeill wrote to me, "We probably could have provided each other with more encouragement for things like studying and not pushing the limits and maybe gotten to graduation with less pain. But maybe we did the most important things with our friendships (you, Fred, Bob, John, Pat) ... with the squadron craziness, ski trips, lake house. After 3-degree year, doing anything else was never a serious consideration."[5]

Even with the strained friendship, the rest of the year zoomed past like a fly-by. The cadet days started fading like the dissipating contrails. We made plans for our lives as officers in the Air Force. We took finals for our last credits, and we readied ourselves for June week with all the festivities that culminated in our graduation and commissioning.

All of us were preparing for new lives, lives in what we called the "real Air Force." Most of us didn't truly know what that "real" looked or felt like, except those who had worn "real" Air Force blue before they arrived at USAFA. The only uniform most had worn, that I had worn, was an academy cadet blue uniform.

Some were preparing too to wear wedding togs. Academy lore had cadets booking the Cadet Chapel for the day of graduation on their first day as cadets, four years ahead, without having even a prospect of a spouse. Certainly, many married under the spires of the silver chapel walls and walked under arched sabers as husband and wife. This year, some of the cadets would be brides. One female classmate ran from the graduation ceremony in the morning to her wedding at the chapel just a few hours later.[6] The press reported the event.

Betsy Joviak Pimentel told me of her bride-cadet dilemma. "When the Class of 1980 got close to graduation there were a group of female cadets who were engaged to be married soon after graduation. The USAFA Officer's Wives Club had a tradition of hosting a Fiancée Tea for all the non-military fiancés and thought it would be a nice gesture to include the female cadets who were about to graduate and get married. The real goal of the tea was to introduce the fiancées to the military and to what the new wives should expect. My wedding was set for two days after graduation, and I received an invitation to the tea. Unfortunately, the day of the tea was the one day I had to spend with my mother after she got into town to finalize wedding details. I sent an RSVP declining the invitation." Apparently, the AOC of Betsy's squadron heard about her response and called her into his office to ask why she wouldn't be attending. He continued to push her to go. "I think he was concerned it would look bad for him if I opted out," she explained. "I said I'd spent 4 years learning what the military would expect of me once I graduated. Finally, I asked him directly if he was ordering me to go. He looked surprised by that but said—no." She didn't go but got a lovely cookbook from one of the spouses with a note saying, "This is all you really missed from the tea. Have a great life!"[7]

During the awards ceremony a few days before graduation, my brother Pat arrived at the cadet field house late and sat amid the cadets instead of with the visitors. When my name was called for an award for Outstanding Squadron Commander, cadets near him commented that I received the prize only because I was female. "She has no right to an Eagle and Fledgling statue," he heard the cadets say. "How do you know?" he asked. "Do you know her?" His flaming red hair should have tipped them off to his relationship to me and to consider their words. "She can't possibly be worthy," they said. "There were only four women to hold that position all year out of eighty. Statistically, alone...." But it wasn't about statistics. I was nominated and met with a series of boards to consider my record and my performance. Cadets were the interviewers. They interviewed cadets from my own squadron. "I am most certain she was the right choice. I know her well," he snapped back. "She is my sister."

Our final day, the day of our academy graduation, started just after midnight. We couldn't wait to become officers. We Firsties of Seventeen decided to hold our swearing-in ceremony at midnight so we could be lieutenants as soon as possible on May 28, 1980. Any active-duty or retired officer could administer the oath of office. Our AOC, Major Henry "Hank" Wilborn, did the honors for anyone without a preferred officer to swear them in. I respected him and would have welcomed him as my sponsor, but even more proudly, I asked my dad to do the honor for me. Dad stood proud, held up his right hand and began the "repeat after me" sequence. "I, state your name." "I, Kathleen Utley," that part I spewed right back. "Do solemnly swear..." and I thought he would pause there as the other officers had done. But with that twinkle of confidence and challenge in his eyes, he marched right on to "to support and defend the Constitution of the United States against all enemies foreign and domestic." And looking him straight in the eyes, I gave those words back to him. "That I take this obligation freely without any mental reservation or purpose of evasion, that I will well and faithfully discharge the duties of the office upon which I am about to enter. So help me God." And I said it ... all of it. We saluted, officer to officer. Then we hugged, father to daughter. Next, that same steely hug that had supported me after I signed the papers to become a cadet drew me tall to officially become a lieutenant of the United States Air Force.

In a brief family ceremony in my room afterward, I presented my parents with a cadet saber framed in cherry and mounted on blue velvet. The inscription read, "Mom and Dad, Thank you. My accomplishments are fruits of your love, 2nd Lieutenant Kathleen Utley, 28 May 1980."

Later the same morning, the Class of '80 marched into Falcon Stadium in our starched parade uniforms. I felt privileged and grateful to be among the almost 900 classmates who marched with me. I did not even feel the high Nehru collar of the cropped parade jacket that normally constricted my neck. My neck elongated with my pulling my head up so tall. This was the final event of a hectic Graduation Week.

The over 4,000 cadets of the cadet wing and guests from all over the globe filled the seats. My parents, both veterans (one Army, one Navy), came to watch the event. My sister, MaryAnn, from Florida came back to see me from closer than the chapel wall this time. My brother Pat had hitchhiked from parts unknown to join the celebration. My sister Barb and good friend Nancy drove from Florida in an MG Midget almost having to bail the car out through the rain-soaked highways of Texas. My high school exchange student "sister" Doris came from Chile, adding the pomp and ceremony of my USAFA graduation to her tour of the United States.

The graduation was exciting. The secretary of the Air Force, Dr. Hans Mark, spoke. He always wrote his speeches himself but asked his staff for ideas for this commencement address. Most thought he should speak about women's role in the military. The auspiciousness of the first graduating class with women seemed to dictate a speech highlighting women. Dr. Mark "decided that the best way to handle the problem was to mention women in the military very prominently somewhere in the speech and leave it at that." He started what he later called his favorite speech addressing the gathered crowd, "General Tallman, Mr. Mayor, parents, families, friends and ladies and gentlemen of the Cadet Wing." The audience erupted with lasting wild cheering and standing applause as soon as the word "ladies" hit the microphone.[8]

After the speeches, the audience quieted as the names of the graduates were called: 899, including 97 women's names. We filed across the stage from left and right alternately, rendering a salute and receiving a handshake and a diploma.

Kathleen M. Conley was the first woman graduate of the Air Force Academy. Her order of merit put her eighth in the class, seven men ahead of her. Reminiscent of day one, the press swarmed around her making her very conscious of every action. She told one reporter that we were all

17. Firstie

sharing this achievement. How generous. The same news reel quoted classmate Terrie Armbruster as saying it took "patience, perseverance and perspiration to get through."[9]

For each graduate, the previous graduated lieutenant waited at the base of the ramp off the stage front and center. He or she rendered the first-ever hard-earned salute to the next graduate. That male cadet, a classmate of mine for the past four years, would not salute me. He looked directly into my eyes turned his back and walked back to his seat. I stood at the top of the ramp shedding my final USAFA tear. *No! It wasn't going to end like this. I wasn't marching down that ramp no matter how many piled up behind me until a classmate saluted me as a lieutenant.* As the next cadet stood confused behind me, another graduate saluted his trailing classmate, walked from the other line across the grass in front of the stage and snapped a professional crisp salute at me. I saluted back, proud to know him, proud to serve with him, so proud and so very ready to be finished.

Once again, the press was ubiquitous. Newspapers, magazines and television reports showed the women of the Class of 1980. Again, it appeared that there were few men in our class. *U.S. News and World Report* of May 26, 1980, featured an article, "Academy Women. Ready to Take Command." Photos showed men at the fringes of the pictures or cut in half, so the women were the focus. The section about the Air Force Academy mentioned "male cadets are still deeply resentful of their (women's) presence here." One male cadet stated, "There is a lot of animosity, and a lot of guys go out of the way to avoid women cadets or speak badly of them." But the director of admissions said, "The Air Force has never gotten 97 women with their qualifications." I was quoted as a "survivor" saying, "we had to go above

C1C (Cadet First Class) Kathleen M. Utley, Squadron Commander, 17th Squadron. Graduation parade, May 1980 (author's personal collection).

and beyond because in the eyes of most men here we started out in the cellar."[10]

June Van Horn graduated number 11 in the class. She recounts that the "press ran up to me for an interview, but I told them I wouldn't talk to them until they talked to all the guys ahead of me. They never came back, and I was good with that."[11]

With the final words of the ceremony that for the first time ever included the word "ladies," "Ladies and Gentlemen of the Class of 1980, you are dismissed," I pulled the hat from my head and swung my arm in a wide whirling circle to toss it high. I let it fly only a foot from my fingertips, though. I snagged it back and tucked it tightly under my arm, safe from all the children released onto the field to retrieve the hats as souvenirs.

That hat meant more than something I could toss away. My sweat, from times I stood on the parade ground as a Doolie, stained its leather interior band a darker color. It waited for me clean and white in the closet while I dirtied my hands and face through the rigors of SERE. I marched in this hat as a member of group staff as a Two Degree. When I was squadron commander, I practiced saber drill for hours to perfect the whipping movement to raise the sharp saber from a lowered salute to a marching upright position. My shoulder was a bruised dark brown for weeks, but this hat saved my ears and probably had stab marks as proof of the many times it did.

The pure Colorado sun filled the bowl of the stadium with heat. The bright blue sky was laced with contrails from the Air Force Thunderbirds flying back and forth in formation overhead. I scanned the crowd for some familial faces. My hair plastered to my head with sweat, I carried the hat up the steps of the stadium to my waiting family. My older sister, back in Colorado four years later, handed me 21 long-stemmed yellow roses, one for each of my years. I enjoyed their beauty there in my arms. I handed her the hat. "This is for little Cathy," my niece who had sat on the wall calling down to me four years before. "She can wear it here sometime or keep it as a symbol and go anywhere she wants to go."

Years later, Cathy gave the hat back to me. She knew how much it would mean to me for my own children to have it. She was the guardian of the hat and some of its history for ten years. Now it hangs on the wall in my home, beside the plaque and saber I had given to my parents. A second plaque and saber hang adjacent, belonging to Fred, the USAFA graduate who shares my life and love. I have grown in love for USAFA too. It challenged me. It showed me my weakest and strongest hours. It taught me discipline and strength and honor.

My daughters or son could have worn the hat someday. They didn't. Mary and Anne chose other amazing universities, and our son, John, as a cadet at USAFA wore his own just like it. I keep it to remind them and to remind me to continue to teach them that there are limits to their behavior, but there are no limits to their dreams. We can all remember that discipline, like exercise, can make you stronger. The hat will be a reminder that a male can raise a family. A woman can fly a plane. Females can be presidents or maybe someday priests. Little efforts can add up to a big difference. Dreams can bring changes. Bring me men can also bring women.

18

Graduate

"It always seems impossible until it is done."
—Nelson Mandela

I thought I could predict who would make a good officer, who would be promoted to general. I still believed that USAFA was a microcosm of the Air Force and a good indicator for career success. I realized later that it, like other universities, was a laboratory. It was a place for trials, trials by fire sometimes. It was the forging process for shaping future leaders and for giving those same soon-to-be leaders the opportunity to refine their understanding of their responsibilities before their true Air Force performance began.

I thought I would be a general. I thought I would spend an entire career in the Air Force, through the silver bars of captaincy, then wearing first gold, then silver oak leaf clusters on my shoulders as a major, then a lieutenant colonel. I imagined myself pinned with silver eagles and, finally, dressed in service blues sporting one star at least. I thought that I would make the Air Force my home and spend my career like my dad serving in the military, moving and remaking myself. It started out that way.

As an officer, I wore my uniform skirt most of the time even though I had resented how it made me stand out at USAFA. Women in uniform were a well-established demographic in the Air Force. There were plenty of female officers on base from other commissioning sources, women who graduated from regular universities and then went to 30 days of Officer Training School to become lieutenants. Other women who wore ROTC blue uniforms on and off throughout their university days now wore the same lieutenant uniform I wore.

I never planned on flying. At five foot two, I wasn't pilot qualified. Peggy, Debbie and June went off to pilot training. Of the 97 women who graduated, 25 were being assigned to pilot training programs.[1] Bob, Ron and Fred went off to learn how to be Air Force pilots too. The number of graduates of 1980, now lieutenants, going to pilot training amounted to

18. Graduate

about 70 percent of the class that year but didn't include me. As a cadet, I almost puked into my oxygen mask during an orientation ride in a T-37. Tall enough or not, I wasn't that interested in flying. I proudly took my computer science degree and went to work to develop software for the planes I would not fly.

I started to spend too much of my lieutenant's pay on phone calls to Fred. We realized that we missed our friendship and our togetherness, and we started to date from 500 miles apart. He was in pilot training in Columbus, Mississippi, and I was stationed at Scott Air Force Base near Saint Louis. We talked often, but we didn't see each other much. There were some amazing dates, though, when he would fly in on a training mission, landing his T-38 Talon in a roar of engine noise right there at my base's landing strip. Those nights we went to the Officers' Club, Fred still in his flight suit, handsome and *Top Gun*-ish.

I could say that USAFA was the best thing to happen to me, but I would be quibbling. During academy days, "quibbling" was the only loophole in the strict "I will not lie, steal or cheat" honor code. Quibbling was not quite lying, a violation that would get a cadet sent to an honor board and possibly kicked out of USAFA. Quibbling was an unintentional lie or incomplete truth and a reprieve from expulsion. Without quibbling, I pronounce that my academy experience was huge in my formation of the woman I have become. Fred, however, is the best thing that ever happened to me, and because of him, Mary, Anne and John. Integrity, service and excellence are part of Fred's makeup without any urging from giant portal lettering. In his next life, he says he is going to be the first man at an all-female institution. Bring Me Women? Be careful what you ask for. It's not always as fun as it sounds, I tell him. He already knows.

Fred and I married just under two years after graduation. He wore his uniform. I chose the traditional white wedding dress. I could have worn my uniform, but that felt like it would throw off my military-life balance. Our wedding was in a small church in the small town I had left to fly off to the academy six years before. Recently, while visiting the academy, my son admonished me for not getting married at the silver-spired icon, the USAFA chapel. He is right. It would have been brilliant with our classmates, male and female, who supported us through USAFA, as our honor guard. Their cadet sabers raised into pointed spires to match the chapel's roof would have been the perfect channel for us to walk through to start our life together.

The Air Force was barely more prepared for the married me than the academy was for the female me. Marrying another service member complicated my position because the Air Force now had to balance two careers with the "needs of the service." Fred's position as a rated or flying officer

Captain Gail Benjamin Colvin (left) and Captain Kathleen Utley Kornahrens, promotion to captain ceremony. May 1984, Rome Air Development Center, Griffiss Air Force Base, New York (author's personal collection).

18. Graduate

outweighed mine as non-rated, so his assignments always took priority. It had nothing to do with gender. We lived apart for a while. I was assigned jobs that were not part of a good career progression for me. I started to know that I could have it all but not all at the same time.

When Mary, the first of our children, made her appearance, it was obvious that the Air Force, two officers, one huge airplane, two careers and a family living together would not all fit in the same plan. I was tired of juggling to keep two career plans in play. I wanted to spend our time building our family and raising our children, not commuting to see each other. I looked at the round, innocent face of our firstborn and knew that the decision to raise her as I was raised, with no limits on her dreams, was worth more than a star on my epaulettes. I didn't think at the time about staying in the reserves to continue to serve, but I was still thinking in absolutes and no one in an Air Force uniform reminded me of the value of that option. Mentors were still missing from the career-life balance equation. The Air Force at that time awarded only four weeks of maternity leave. That was the final sign that I needed to choose motherhood over captaincy if the Air Force wouldn't help with the balance. I donated my uniforms and walked away.

After my military service, I chose community service because that is what fills my heart with the greatest feeling of success. When I was young, my father spent hours reading to the blind couple who lived down the street from our Florida house. My mother was also a community server and a gatherer of the lost. Many Sundays, she'd bring a stray parishioner home from church to participate in our family pancake feast to talk politics, social justice and religion, essential parts of our home conversations. Military and community service were our family cornerstones. When it came time to choose my second career, I chose the second type of service from which I was raised.

Through the years, I have tried to be the mentor to young hearts that my father was to me. To my own three children, of course, but also to the young of my community who need an extra boost in believing in themselves and their possibilities. If they are lost in math, we talk about fractions, but I make an effort to add little comments about their artistic handwriting or the lovely timbre of their singing voice. I want them to know that their gifts are beyond the measure of grades and IQs. I learn so much from them that I feel most of the pluses of the encounters weigh in on my side of the balance. I tutor adults too, immigrants from Somalia and Mexico who are trying to find their place in America as I remember trying to find mine at USAFA.

* * *

I saved a bit of memorabilia from my cadet days. My blue *Contrails* with its torn spine and loose pages has a place of honor on my desk. The plaque and saber I gave my father is back on the wall of my office. I still have the hat that I didn't toss at graduation. I have my parka in the back of my closet—too warm for most days and I haven't needed a Playboy Bunny outfit again since Doolie year. The magnets and playing cards with the picture of the "Bring Me Men…" ramp that I bought at the visitor's center when our family visited USAFA in 2003 are stored away in a filing cabinet.

The words "Bring Me Men" were removed from the academy right before that visit. The elimination of those words that marked the ramp during my years at USAFA and for more than 20 years after was a result of the academy's very public sexual assault scandal. This scandal was not during our academy years, but it is a disappointing part of the history of women at USAFA that began with us 27 years earlier.

The press interfered and brought attention to us which made our assimilating into the cadet wing more difficult. Later, the same press were the ones that brought the stories of others who hadn't felt heard to the conversation. Coverage of the sexual assault scandal seemed more widespread coverage than what the newspapers had provided on the arrival or the graduation of our class.

In May 1994, the *New York Times* headline read, "Air Force Academy Zooms in on Sexual Harassment." "Since February 1993," the article stated, "when a female freshman at the Air Force Academy here told the campus authorities that several young men had sexually assaulted her outside the cadet gymnasium, a dozen other women have stepped forward to lodge complaints."[2]

General Hosmer, superintendent of the academy in 1993, met soon after the report of the assault with "most of the Academy's 518 female cadets." Fifty percent of the women knew of some case of sexual harassment. When he later met with the male cadets, only 9 percent were aware of sexual harassment cases. There was very disparate awareness of the problem. "The bigger problem," Hosmer was quoted, "was the climate here."[3]

USAFA instituted sensitivity training for cadets and staff and the establishment of a Center for Character Development at USAFA to bring ethics, honor code issues and human relations training to a centralized department. The academy also set up a rape crisis hotline and implemented confidential reporting with an amnesty policy.[4]

Then in 2002, two female cadets sent an email to various news media organizations and to the office of Colorado senator Wayne Allard. The *Air Force Times* reported their claim of having "been assaulted at the academy and largely ignored when they complained about it."[5] Another article came out in the Denver weekly *Westworld* about assaults at USAFA. Other women came forward with their stories. A Colorado Springs rape crisis center reported that in the 15 years prior, 22 cadets sought confidential help.[6] Allard's office said it had heard from at least 20 women, including current cadets. There was also a troubling secondary claim: "because they came forward with complaints, these women were themselves punished for infractions such as fraternization and underage drinking, even while their alleged assailants received lesser punishments or escaped sanction altogether."[7]

Virginia senator John Warner stated that "some facts give rise to the conclusion that a climate existed that was actually hostile to female cadets. Some facts provide a basis to support a conclusion that the promise of a safe and secure living and working environment for female cadets ... was undermined."[8]

The secretary of the Air Force directed the Air Force inspector general to review these accusations and to investigate cadet complaints concerning the alleged mishandling of sexual assault cases. The secretary of defense appointed an independent seven-member commission led by former congresswoman Tillie Fowler to review the academy's policy on sexual harassment. The Fowler Commission proposed 21 recommendations from greater access for women cadets to rape crisis centers and counseling to increased oversight at the academy by Headquarters Air Force and the Academy Board of Visitors.[9] The academy implemented the "Agenda for Change" on March 26, 2003.

Gail, who was in the role as vice commandant of USAFA in 2007, responded when asked why reported sexual assault and harassment cases had increased that "cadets have trust in the system and view it as a safe place where they can seek help." "The challenge ahead," she said, "is for the academies to continue advancing programs to prevent sexual assault and sexual harassment from happening in the first place. It's about creating a culture and climate of respect, both for oneself and others."[10]

* * *

During my four years at the academy, how many women were harassed?[11] All. I can say with almost complete certainty that every woman in my class experienced gender-based harassment of some kind. Everyone, men and women, was harassed. But the women cadets experienced something different. We were harassed ... because we were women ... in a male-dominated environment. How many were assaulted or touched without permission even with the established "May I touch you?" requirement? I do not know. I have since read in many articles and heard in much congressional testimony that even *one is too many*. I completely agree.

One of my classmates remarked, "It mirrors what happens on college campuses across the United States. The disappointment lies in the hope that the service academies attract a different type of young person. Higher standards are advertised but not necessarily produced."

The "Bring Me Men..." sign did not come down before I arrived at the academy. I did not see the sign as an affront or as a symbol of my being less. My life at the time was more about basic survival than about institutional changes that were beyond the already enormous change of our class of women invading that male domain. My priorities were about completing the requirements to the best of my abilities and graduating to prove the capabilities of our gender, the correctness of the decision to integrate.

There was talk of removing the sign in the planning for our arrival in 1975. The academy decided to follow common English understanding of "men" in the generic sense. They also cited concerns about making changes that were detrimental to academy tradition, would be high cost and concerned that "changes should promote positive attitudes toward women cadets."[12] The hostility that the cadets and graduates would have felt if the sign had changed as we arrived would certainly have been directed at us. The issue with the sign was on the same page as hairstyles in the annual academy history. "Literally hundreds of problems had to be resolved ranging from what to do about the 'Bring Me Men...' inscription at the portal to Vandenberg Hall to what style haircuts would be given to women cadets. The slogan remains unchanged yet, but the haircut problem was solved...."[13]

There was also discussion to remove the sign in 1999, but the academy concluded that the sign should remain for historical reasons.

When I read about the sexual assault scandal that brought attention again to the sign in 2003, there were so many instances of women cadets who were ostracized, criticized and punished because of the flaws of the institution that their situations exposed—the flaws of the culture, the mistakes of the reporting systems, the accepted inuendo of the humor, slogans or patches. Had the sign been removed in 1975 before our arrival, it would have been a change about gender, and the backlash would have made our integration much more difficult. In 2003, the removal of the sign seemed more about a reflection on a necessary change in the culture at USAFA and of the character expectations of a USAFA graduate.

The USAFA Association of Graduates was tasked to facilitate choosing the new words. "We are accepting all ideas," a graduates' spokesman said. Air Force Secretary James Roche and Chief of Staff General John

18. Graduate

Jumper "mandated any slogan must more suitably represent the aspirations of the entire cadet wing and the core values of the Air Force."[14]

The sign's ultimate removal resonated strongly with members of our class, male and female, and with many graduates of the academy. Men and women argued the sign represented tradition at the academy and had nothing to do with sexual assaults. Others, also men and women, remarked that it was time for the exclusionary words over the portal to be replaced.

At first, I was shocked and saddened to see the sign, a piece of my historical reference at the academy, gone, but my understanding changed with the very wise words of my daughter Anne. As her sister Mary had asked me when we saw only the outlines of the missing words, "Why would you care, Mom?" Anne told me later, "They are just words." She made me realize that more important than the words were the changes happening in the culture of USAFA. Changes to ensure that all cadets felt included, all were represented and all felt safe and supported. Those changes were the bigger issues, bigger than the giant silver words. "Plus," Anne explained, "removing the words left space for the next generation to choose their inspiring words."

My assessment also changed when a female classmate proposed that we each knew how we interpreted the words of the sign. What we did not know is how some others interpreted the words. Maybe some male cadets without the right understanding of the principles and character expectations of the academy for leaders of our nation's military, without the appreciation of human dignity in any setting, might have read into those words other malicious, inequitable, exclusionary meanings.

The sign over the ramp cannot be misinterpreted now. The ramp is called the Core Values ramp and reads, "Integrity first, Service before self, Excellence in all we do."

In 2012, my son, John, arrived at the base of the Core Values ramp. He stood where I had stood, where Fred had stood with an expectation of a demanding four years of training, character building and education. He expected his class to include women. It did: 236 women who stood at the base of the Core Values ramp with John that day.

I became the letter writer as John was going through the fires of Basic Training. I accidentally ended up in the dusty paths of Jacks Valley on a tour at the exact moment when my son came stumbling over a dirty ridge and through a cloud of red smoke on the assault course. The cadre still wore their USAFA T-shirts and fatigue pants. My heart pounded with memories and to see him there. He stumbled, fell and a large assault course cadre member grabbed John's rifle trying to wrestle it from him. John held strong. His face was caked with mud, and he could barely raise his boots from the dust. We couldn't speak to each other. He saw me at a distance. I saw him and half-raised a hand to signal hello.

I wrote to him that night.

> John,
>
> *Seeing you today was so hard, but I was so proud of you. You are just what USAFA is all about: character, strength, excellence. I didn't want to upset you. I know it was hard to have me there during such a tough time, but I was so sure you could do it. You looked strong. A warrior.*
>
> *Love lots,*
> *Mom*
> *NIC*

I mailed him some Latin from my father, some words of encouragement and hopefully some relief.

John graduated in 2016, and we presented him with a broken piece of marble from the terrazzo rescued from a scrap pile with his name inscribed in the white stone. Gail was his strategic mentor during his years at USAFA, not baking cookies or having him over to do his wash but offering him valuable perspective on the academy and life, much like what she did for me.

* * *

Our class was honored as "Architects of Transformation" at our 30-year reunion. The academy recognized the men and women of our class as members of the historic class. But we were not the architects.

In 1976, it seemed to me that the academy was ill-prepared for the arrival of women. The more I learned, the more I knew how prepared they were. Those who planned and strategized years before we arrived at the "Bring Me Men…" ramp, who ordered books for the library, who stood in as surrogates, who whispered encouragement and who accepted skirts into the uniforms of the long blue line, they were the architects of the transformation at USAFA. We were often accidental pioneers who arrived after the architects formulated the plan, and we fought to bring their blueprint into reality. An academy historian once called our class "grace over steel." I am proud to be one of such an amazing group.

In 2016, 40 years of women at USAFA were celebrated by inviting 1980 graduate women to stand on the football field at halftime. My son, John, was then a recent USAFA graduate still at USAFA, so I flew to Colorado to participate as the bonus to a visit with John. Marianne Owens LaRivee made posters spelling out the numbers and letters of "80's LADIES" for each of us to hold. We wore our old parkas, if we still had them. Gail Benjamin Colvin wore the parka of her son James who graduated in 2012.

18. Graduate

Marianne still had and wore her dress cap, not the felt mushroom but its somewhat better-looking replacement that we wore while we were still cadets. Janet Libby Wolfenbarger stood in the line with us, as smiling and positive as I remembered her when we were both still cadets. Tanya Senz Reagan, Alene Dowden Saleck, Lorrie Morse Kresge and Bonnie Schaefer Schwartz held white letters on blue posters too. We stood on the field holding our signs and received cheers and applause, loud and sincere. Walking back up the stands, I felt like a rock star. People screamed, "Great job, ladies" from their seats. People stood and saluted as we returned to our seats. Some came up to us and shook our hands, saying, "Thanks for what you did."

Why hadn't the time at USAFA been a supportive time like that? But that thought can only last a minute because it was the forging in the fire of the challenges, the taunts, the yelling of exactly opposite sentiments that built our individual strength, our strength as a group, our resolve to go beyond the minimum, to build a rock-steady foundation for the women who would come after us. Easy times do not beget strength. Strong muscles come from lifting weight, strong hearts come from heavy workouts, strong minds come from stretching your brain to the absolute endurance of your sanity. Do not think that we were superwomen, superhuman. We

In 2016, celebrating 40 years of women at USAFA, Class of '80 women were honored at an AF Falcon football game. It was a memorable night. 80's Ladies from left: **Gail Benjamin Colvin, Bonnie Schaefer Schwartz, Allene Dowden Saleck, Kathleen Utley Kornahrens, Marianne Owens LaRivee, Lorrie Morse Kresge, Tanya Senz Regan, Janet Libby Wolfenbarger** (author's personal collection).

failed. We cried and limped and slumped in our uniforms until the day we graduated when we stood tall stretching the fabric of the parade uniforms tight from chin to shoe top.

I am so thankful that I was a member of that 1976 group of women pioneers. I got in on a waitlist. I cried and dropped out of pounding runs. I made leadership mistakes and took the heat for some mistakes that the leadership of USAFA made. Still, I eventually led a group of over 100 cadets. I temporarily lost my hair, my individuality and my family on my way to finding confidence, independence and strength. I bonded forever with a group of incredible and brave women. I bonded forever with one of the men as my husband and with many of the other men as lifelong classmates and fellow Air Force Falcons. I grew in respect for my father, mother and brother who served. I learned to appreciate those who choose not to serve and those who protest the causes for which others fight. I learned the value of discipline from saying yes when your heart is screaming no, to saying no and thinking beyond the regulations and rules that seem to indicate a single course of action. I know that a woman's place is where she can grow, learn and thrive. I learned that a casual comment of possibility from my high school teacher or to a tutored student can take root and blossom. I know that my small signature with my father's black military pen made me a new future. I realize that an individual contribution to a large movement can change outcomes. I spent four years marching up and down the "Bring Me Men..." ramp, my heart pounding less each time. "Bring Me Men" brought the United States Air Force Academy women and brought the women, including me, a challenge and an accomplishment of a lifetime.

Ultimately, no matter what the words over the portals read, we should read them as "Bring Me All."

Epilogue

> Never doubt that a small group of thoughtful committed citizens can change the world.
> Indeed, it is the only thing that ever has.
> —Margaret Mead

When I went to our tenth reunion in 1990, I had to buy a suit, as I was no longer an Air Force officer. A smattering of classmates now wore mufti among the active-duty members still wearing the blue uniform for the official photo. I remember Peggy coming into the evening's formal dance in a black-and-white dress that fit her gorgeously. She was out of the Air Force and had just been hired by United Airlines. Her smile, still beautiful, jumped right onto her face without coaxing. I was so happy for her to have her dream of flying and her family. She had given up on the astronaut track but not on reaching for stars. She still flies and still smiles in an instant. We too have a bond of the "sisters of the brotherhood" and great memories of some crazy times at USAFA.

After USAFA, Debbie Wilcock Ziebart went to Williams Air Force Base for Undergraduate Pilot Training, first as a student, then as instructor pilot to teach other student pilots in the T-37. In her next assignment, she flew the KC-135 refueling airplanes for the Air Force and, after leaving the Air Force, flew the remainder of her career with FedEx.[1]

Gail Benjamin Colvin was at that first reunion with a beautiful brand-new baby, James. She was just beginning her struggle to balance family and force. She was a perfect mother, and the love reflected in her eyes for that boy was beyond stars. I wonder what she sang to him at night *... used to wear those high-heeled shoes, now I'm sporting combat boots....* She had another son, Jordan. She stayed in the Air Force and ran from the Pentagon on September 11, 2001, praying while she ran to be able to see her boys again. That first son is now a USAFA graduate.

June Van Horn Lindner flew after graduation. When she moved to Dyess Air Force Base in the KC-135, she was the only female aircraft commander there. "Before I even arrived, a copilot had informed the training officer that he would never fly with me. Ever. Not having a female aircraft commander and a male copilot. Not happening!" She won him over with her skills, her pranks and her friendship, not unlike what she did to earn respect at the academy. In the real Air Force, she also felt "for many years, women were not allowed to be average. If we messed up, even a little bit, the attitude was: typical female—can't fly." She left the Air Force. "I just didn't feel I could give 100% to the Air Force and 100% to my family.... For me, it was the right, but agonizing choice, to get out," she stated. She chose the challenge of homeschooling her children and supporting many charities particularly helping victims of natural disasters dig out of their damaged homes with incredible selfless people from all over the United States. She says by the end of the work, "we have eternal bonds with each other, similar to the bonds we share with all of our awesome classmates from USAFA."[2]

I admire Marianne Owens LaRivee's service wearing her Air Force blue uniform with a special pin with the scales of justice over her name tag. Marianne explains, "After the Academy, I went to Duke University School of Law and spent the rest of my 21 years in the Air Force as a Judge

The 9th Squadron women in the Cadet Chapel at their 20th reunion. From left: Kathleen Utley Kornahrens, Gail Benjamin Colvin, Peggy Walker Bertaina, Debbie Wilcock Ziebart, Marianne Owens LaRivee, June Van Horn Lindner (author's personal collection).

Advocate General (JAG)." She married a classmate, Dave LaRivee, has two children and has been stationed nine different places in the United States and overseas. After she retired, she taught undergraduate law at the University of Colorado in Colorado Springs for 15 years.[3]

* * *

Our classmate Susan Helms was our first major star. She ventured closer than any of us to the celestial bodies on three different NASA missions. Susan explains about her time at USAFA: "That first year [at USAFA] was an interesting experience. For me personally, the thing that struck me was that I had never been in a male environment like that before. When I came to the Air Force Academy, I was under the impression that guys usually liked having women around them in college, so it was quite a shock to me to see that there was some resentment along that line." She added, "I think we all figured out real time how to develop these coping skills for living in one kind of culture and, basically overnight, turning it into a different kind of culture."[4]

Telling about academy experiences that were helpful in her Air Force career, Susan said, "There is one thing that I never guessed ahead of time. When I was a cadet, of course the unique thing about the Class of 1980 was that we were newly integrated in a male dominated environment ... eighteen years later, I got selected for a flight assignment as a crew member in the Russian space program.... The thing I drew on more than anything else in my previous experience was the adaptation process of being in the first class of women at the Air Force Academy. That ended up serving me very well eighteen years later as I learned how to adapt as a woman in the Russian space culture. If gender ends up being a reason to question competence, that could be bad. So you always try to make sure you are as competent as possible, not as a woman but as a crewmember and meeting the standards for all crew members."[5]

The "80's Ladies" were invited to her space shuttle launches in the early '90s. Arriving at the viewing area at 0430, we parked our cars and wandered the dark Cape Canaveral causeway searching for each other. Peggy, Debbie and I and a group of classmates, male and female, stood in awe when that rocket rumble shook the ground and the roar of the booster engines reached us moments later. One of us was not only heading to the heights, but she wore an "80's Ladies" shirt for some publicity photos and zero-gravity television spots. I was not jealous. My admiration left no room for any other emotion. Susan did the academy and the women so proud. I thought of myself as one tiny stitch on the "80's Ladies" logo sewn into her Air Force blue shirt.

Janet (Libby) Wolfenbarger wore the most stars of any 80's Lady. The Senate confirmed Air Force lieutenant general Janet Wolfenbarger for promotion March 26, 2012, making her the first female four-star general in Air Force history. She was the commander, Air Materiel Command, Wright-Patterson Air Force Base, Ohio, leading 80,000 people and managing $60 billion annually.[6]

When speaking about her experience at the academy, Janet remarked, "Some male cadets in the Cadet Wing firmly believed that Congress was wrong to allow women to attend their previously all-male institution. A subset of those individuals made it their calling to prove their point. In hindsight, I think these cadets feared that the institutional standards would somehow have to be lowered because of women becoming a part of the Cadet Wing, and their resulting experience would then be diminished." She expounded that we spent four years proving them wrong by not only surviving but thriving in the very challenging environment.[7]

She credits her dad, an active-duty Army major, with sage advice about what to expect when she arrived in 1976. "They're going to strip you of all your rights and then hand them back to you one at a time." She acknowledged, "I didn't fully understand what he meant at the time, but I quickly learned that he knew what he was talking about. The Academy placed me into situations that really challenged and stretched me in any dimension you can envision—mentally, emotionally, academically, physically." She said it took some time before she realized "the value of that crucible experience, … that as I came out on the other side of those challenges, … I was far more capable than I ever thought I could be."[8]

I was fortunate enough to be included in the retirement ceremony for General Janet (Libby) Wolfenbarger, a classmate and friend. It was a wonderful day with a moving ceremony in a room filled with stars on the shoulders of many Air Force uniforms. General Welsh, chief of staff of the Air Force, spoke about General Wolfenbarger and her Air Force career beginning with her time at the academy with the Class of '80. "Will the 80's Ladies that are in the audience please stand and be recognized?" he asked. I stood along with about eight other female classmates who were in attendance. I was humbled and a little embarrassed to be standing and recognized in the midst of all those stars. The secretary of the Air Force, Deborah Lee James, when speaking about Janet's history, asked the "Ladies of the '80s" to stand. We stood. No one corrected her, but we smiled. How amazing to have a woman in that office talking about a woman who was the first four-star general.

During the evening festivities, Janet gave a speech. Once again, the "80's Ladies" were asked to stand and be recognized, but then Janet also recognized the male classmates who were present, including a four-star general and an industry CEO. There were amazing men in our class too.

Finally, Janet remarked, "When we were called ladies it had a different timbre. It was not so positive, but we turned it into a positive."

There are scores of stellar 80's Ladies. I was proud of Linda Garcia Cubero's success. She was inducted into the National Hispanic Engineering Hall of Fame and named one of the "100 Most Influential Hispanics in the United States."

In an article in *Latina Style* magazine, she talks about her decision to go to the Air Force Academy: "As the only Hispanic woman to graduate that year from any of the academies, I've faced challenges as a woman and as a Latina (my father is Mexican-American and my mother is Puerto Rican). So why did I put myself through that 'pain'? I wanted to follow my father's footsteps as an AF officer, serve my country, travel, get the best education possible, and take advantage of the unique opportunities available, like flying Cessnas, soaring in gliders, and earning my free-fall parachute wings."

"I love a challenge," she continued, "so when I was told by my guidance counselors that I wasn't good enough to make it into USAFA, I set out to prove them wrong. I was ranked 25/485 students in my class and was a member of the National Honor Society, yet they told me I was 'too dumb' to get in. My motto is, 'You tell me I can't, and I'll show you I will!'" She proudly served on the Air Force Academy Board of Visitors for four years.[9]

Bonnie Houchen not only served but also worked to support other women to be firsts. After 17 years in the Air Force in space and missile operations, she "declined the rank of lieutenant colonel and senior service school to avoid the threat of 'Don't Ask, Don't Tell.'" However, her service continued. She became the assistant commandant for coeducation for women at The Citadel, the Military College of South Carolina as principal adviser on female issues from 1997 to 2000. Women were first admitted to The Citadel in 1996. She is actively involved on the board of Zoomies (term for cadets or USAFA graduates) against Sexual Assault.[10]

Along with Janet Libby Wolfenbarger, two other women had been squadron commanders along with me during our Firstie year. I didn't know them well then, but I knew of their standing in their squadrons, of their reputation as sharp cadets.

When Betsy Joviak was a high school senior from rural Ohio, Air Force Academy cadets came to her high school for a grassroots recruitment program. It sounded intriguing to her with the mix of academics, leadership and athletics that matched her interests and record. She sent in the postcard they left for interested applicants, finally got the right socks and studied and played basketball and volleyball throughout her cadet years.

In her Air Force career, Colonel Betsy Joviak Pimentel supported Air

Force Space systems in planning, development and testing. She later told me that the ATO program was like the academy fit checking for the women's arrival. Fit checking, like on a space craft, is something she would know much about. She commanded a missile squadron and was a director of defense policy on the National Security Council. Following a 20-year career in the Air Force, Betsy joined Northrop Grumman/TASC and then Stellar Solutions Inc. as a vice president for defense programs.

Sue Desjardins marched as her squadron commander too, and although I did not know her well then, I saw her across the terrazzo as a natural leader. She looked perfect in the role: tall, poised, serious but not rigid. She exuded leadership and poise. She had saved waitressing and babysitting money in high school to pay for college, but the opening of the academies to women gave her another choice. "My father suggested, with my mom's concurrence—or trepidation, one or the other—in his words: 'Let's see if we can get in. Let's try it,'" Desjardins later recalled. "I got an appointment (to USAFA) in April of my senior year, and about five days after graduating from Portsmouth High School I was on an airplane headed to Colorado Springs. I had never been west of New York, and that was my second airplane ride ever. That's how it all started."

Sue Desjardins as a general did not surprise me. She also became the first female commandant of cadets at our Air Force Academy.

"Integration of women into the military has not been without issues, prejudices, mistakes and challenges along the way … many of which I can honestly say, I have experienced at some point over nearly 40 years," General Desjardins expressed. "But as these changes in the military were occurring, so too was society changing, or perhaps it was the other way around. The progress and the process continue, and the challenges are being met and, in many cases, overcome."

An embodiment of professionalism and discipline, she motivated the cadets during a time of war. She was praised at her departure for "helping prepare thousands of cadets for war and restoring the Academy's reputation after a series of scandals that plagued her two predecessors." She led by example, running along with cadets, "leading the tradition of doing a push-up for each point the Falcons score." After her three years as commandant, she said, "It renewed my faith in a lot of things. Especially in our future."[11]

Sue also told of the importance of mentors in her career. She served under several commanders who made a point of mentoring her and not just using her for tokenism. "They would suggest to women, 'You should go for this or try this or would you like to compete for this?' Had it not been suggested to me, maybe I wouldn't have done it," she explained.[12]

Her vice commandant was none other than my pugil stick opponent and dear friend Colonel Gail Benjamin Colvin. Her title was USAFA Vice Commandant of Culture and Climate. She wore her skirt on Tuesdays in solidarity with the female cadets. I still think of her as fun and flying above the fray, but I know she can focus on the core of a problem and navigate through the flak.

Colonel Colvin worked to understand what it means to make sure that people feel included, that everyone get equitable treatment, that everyone is valued. Gail relates, "It really opened my eyes to what are the experiences our cadets are having and how do we help develop into leaders of a very diverse force."[13]

Gail worked for five superintendents as the director of staff. She was selected as a distinguished graduate of the U.S. Air Force Academy in 2021. In her preview for that honor, she explains "that diversity is really a part of who we are as leaders of character. It's about integrity. It's about respect and dignity for others. And if we are developing people to lead in diverse forces they need cultural competencies, they need to be able to lead others who don't look like them, think like them or act like them and be able to appreciate and value everyone's contribution."[14]

My female classmates soared to astonishing heights, figuratively and literally. As a general, an astronaut, a pilot, a community servant or an Air Force "lifer," they certainly achieved success and influence, especially in the Air Force. Gail calls the 80's Ladies "my jewels in my jewel box whose inspiration and example has been a source of strength for me throughout my entire career." I know we are all that for each other. The 80's Ladies seem to have grown closer since graduation as we came to realize that momentousness of our shared achievement as the first USAFA women graduates. We get together as a group after each class reunion.

The men achieved great heights too. We honored the men of the Class of '80 at one reunion. They have been stalwart in their support of their female classmates. Within the military metrics, our class has excelled: there were 438 pilots from our class, 62 navigators, battle managers and combat controllers. We claim an astronaut (Go, Susan), Thunderbird pilots and two four-star generals (Janet Wolfenbarger and Paul Selva), 29 other generals (brigadier, major and lieutenant generals). Classmates served in Grenada, Panama, Kosovo, Iraq and Afghanistan and earned hundreds of medals. Many served their whole careers in the service of our country.

So many of the men and women of my class were successful and not always in uniform. Many, like myself, found ways to bring integrity and honor to other service, not only the military type. Our class includes 22 PhDs, 13 MDs, 13 attorneys, a state representative, a state senator and at least one CEO. Some chose community service, others service to their families or serving in industry or education. Graduates from USAFA 1980 work for the Air Force, the State Department, the Department of Energy and the Secret Service.

* * *

One time when I was out at the academy, I met some of the then-current amazing female cadets, maybe one day to be generals but certainly one day to make their marks on the world. During a tour of the academic building with my son, John, we met a first-class cadet explaining her sculpture on display. She was heading for a master's degree in international affairs after graduation. Assuming I was a graduate, she asked me my class year. "1980," I responded. "Oh, okay," she said, seemingly unimpressed. John was astounded that she didn't comment about my being part of that first class with women. As much as I want our battles to be remembered so that others will have hope in their struggle for a place, a part of me was happy that she assumed a female to be an alumna and not recognizing the significance of the year, maybe thinking that women had always belonged at USAFA.

We lunched with the cadets and there met some of the outstanding young male and female cadets who were leading the entire wing that semester. One female cadet was graduating with a bachelor of science in biology with three foreign language minors. She was later named top graduate of 2008, and she became the first Puerto Rican and the 35th academy cadet to be awarded a Rhodes Scholarship. Amazing and awe inspiring in any environment, her accomplishments seemed like the dreamlike culmination of the path started in 1976 with the first female steps marching up the "Bring Me Men…" ramp.

At every football game, the Air Force crowd stands and sings the first part of the Air Force song. It begins, "Off we go, into the wild, blue yonder." We sang it too at football games, on our feet and clapping enthusiastically to the beat. We stood for its playing at many parades and other official functions. It still straightens my spine. But the third verse of the Air Force song is the bridge where the tempo changes and the words don't encourage clapping. My singing of that part of the song is halting with tiny choking squeaks when the emotion keeps my voice from the words.

As tradition, after a game, the football team always removes their helmets, hold hands and stand in front of the student section of their fellow cadets. The team, the cadets and many grads somberly sing that third verse. "Here's a toast to the host / Of those who love the vastness of the sky / To a friend we send a message of his brother men who fly." Well, not those exact words, not anymore. We sang those words in 1980, but in 2019, then–Air Force chief of staff, General David Goldfein, ordered the words be adjusted to reflect and recognize the women in the Air Force after he stood with the USAFA women's volleyball team and sang the old words: "his brother" words. The words of the verse are now "To a friend we send a message of the brave who serve on high." "These new lyrics speak more accurately to all we do, all that we are and all that we strive to be as a profession of arms. They add proper respect and recognition to everyone who serves and who has served," Goldfein said.[15] Another story of changing words and hopefully attitudes for the better.

The original ATOs of USAFA have a silver pitcher, the ATO jug, in a display case in the back of Doolittle Hall. All the names of the original ATOs, including Colonel Hess who wrote the plan for our arrival, are engraved in the silver jug. There are 134 names for the three years of ATOs who provided leadership and mentoring for the brand-new USAFA and its cadets. The jug is filled usually with scotch, Bob Hess told me, and the ATOs who are together pour their drinks from the jug. The recipient of the last full measure from the jug must provide the refill.[16]

We women of the Class of '80 have thought to commission a jug, a carafe, a keg, something big enough to raise a toast to the men of our class who sang along with us, to others who came before who planned for our arrival from bras to latrines, to all 157 women who came to the base of the "Bring Me Men…" ramp on June 28, 1976, to begin the journey to integrate women to those men, and to toast to each other the 97 women who completed that journey.

Chapter Notes

Prologue

1. "The Coming American by Sam Walter Foss," The Bard on the Hill, December 2, 2011, https://thebardonthehill.wordpress.com/2011/12/02/the-coming-american-by-sam-walter-foss/.

Chapter 1

1. Grace Lichtenstein, "Sex Barrier Falls as Women Enter Air Force Academy," *New York Times*, June 29, 1976, https://www.nytimes.com/1976/06/29/archives/-sex-barrier-falls-as-women-enter-air-force-academy.html.
2. George V. Fagan, *The Air Force Academy: An Illustrated History* (Boulder: Johnson, 1980), p. 190.
3. Gary Watson, "Peter, Paul and Mary Tell Jone French Story," *The Free Press Second Front Page*, April 13, 1970.
4. E.A. Muenger, "Fact Sheet on the Integration of Women at the Air Force Academy," edited by Joan Varra (Colorado Springs: United States Air Force Academy, March 1977), p. 1.
5. William J. Wallisch, "The Admission and Integration of Women into the United States Air Force Academy," Dissertation, University of Southern California, 1977, p. 81–82.
6. "Sex Discrimination: Overview of the Law," Home, U.S. Department of Education (ED), February 16, 2022, https://www2.ed.gov/policy/rights/guid/ocr/sexoverview.html.
7. "Exemptions from Title IX," Home, U.S. Department of Education (ED), March 8, 2021, https://www2.ed.gov/about/offices/list/ocr/docs/t9-rel-exempt/index.html.
8. William J. Wallisch, "The Admission and Integration of Women into the United States Air Force Academy," Dissertation, University of Southern California, 1977, p. 83.
9. Muenger, "Fact Sheet on the Integration of Women at the Air Force Academy," p. 1.
10. Wallisch, "The Admission and Integration of Women," p. 102, 104.
11. Donald M. Fraser, Fortney H. Scott, and Jacqueline Cochran, *Hearing Before Armed Services Committee, Concerning the Admission of Women to the Service Academies*, H.R. 10705 ed. (Washington, D.C.: U.S. Congress, 1974), pp. 491–492.
12. Wallisch, "The Admission and Integration of Women," p. 104.
13. Henry S. Fellerman, *History of the United States Air Force Academy*, 1 July 1974–30 June 1975 ed. (Colorado Springs: United States Air Force Academy, 1976), p. 13, 14.
14. The Editors of Encyclopaedia Britannica, "WAVES," *Encyclopedia Britannica*, September 27, 2013, https://www.britannica.com/topic/WAVES-United-States-naval-organization.
15. United States Air Force Academy, *USAFA Catalog 1975–1976, Addendum for Women Admissions Procedures* (Colorado Springs: USAFA, 1975).
16. United States Air Force Academy, *USAFA Catalog 1975–1976*.
17. "Commandant's Committee on the Integration of Women into the Cadet Wing—Minutes Meeting #7" (Colorado Springs: USAFA, September 18, 1975), p. 3.
18. Fellerman, *History of the United States Air Force Academy*, 1 July 1974–30 June 1975 ed., p. 14.

19. Henry S. Fellerman, *History of the United States Air Force Academy*, 1 January–31 December 1976 ed. (Colorado Springs: United States Air Force Academy, 1977), p. 19.

Chapter 2

1. George V. Fagan, *The Air Force Academy: An Illustrated History* (Boulder: Johnson, 1988), p. 3.
2. Fagan. *The Air Force Academy*, pp. 32–33.
3. Fagan, *The Air Force Academy*, p. 81.
4. Henry S. Fellerman, *History of the United States Air Force Academy*,1 January–31 December 1976 ed. (Colorado Springs: United States Air Force Academy, 1977), p. 271.
5. James P. McCarthy, U.S. Air Force Academy Oral History Interview, Other no. 123, February 18, 1977. Clark Special Collections, USAFA, p. 7
6. Fellerman, *History of the United States Air Force Academy*, 1 January–31 December 1976 ed., p. 118.
7. Gail Colvin, U.S. Air Force Academy Oral History Interview, Other no. 1042, September 15, 2010. Clark Special Collections, USAFA, p. 6.
8. *Contrails: The Air Force Cadet Handbook* 22, 1976–1977 ed. (Colorado Springs: United States Air Force Academy, 1976), p. vii.
9. Dennis A. Williams and Martin Kasindorf, "The Sexes: Bring Me People," *Newsweek*, July 12, 1976, p. 24.
10. McCarthy. U.S. Air Force Academy Oral History Interview, p. 17.
11. Lisa Underwood, Letter to Kathleen Utley Kornahrens, *"Shady Rest" Re: Question to 80's Ladies*, March 18, 2003.
12. Fagan, *The Air Force Academy*, p. 78.
13. Fagan, *The Air Force Academy*, p. 191.
14. Betsy Joviak Pimentel, Letter to Kathleen Utley Kornahrens, *Re: Publishing*, February 22, 2022.
15. Colonel Robert Hess (Ret.), Plans for Admission of Women, Personal, February 24, 2016.
16. Stanley C. Beck and Walter R. Berg, U.S. Air Force Oral History Program, Other no. 422, January 14, 1988. Clark Special Collections, USAFA, p. 501.

Chapter 3

1. The Air Force Cadet Wing (USAFA), "Air Force Cadet Regulation 50–8, Attachment 3," Colorado Springs: USAF Academy, December 22, 1976, p. 24.
2. Marianne Owens LaRivee, Letter to Kathleen Utley Kornahrens, *Bring Me Men Manuscript*, February 27, 2022.
3. Gail Colvin, U.S. Air Force Academy Oral History Interview, Other no. 1042, September 15, 2010. Clark Special Collections, USAFA, p. 16.
4. "AOG Service Spotlight: Gail Colvin: U.S. Air Force Academy AOG & Foundation," August 2021, https://www.usafa.org/Service/SpotlightAugust2021.
5. John B. Taylor, "The New Cadets," *Airman* XX, no. 12, December 1976, p. 5.
6. Colonel Robert Hess (Ret.), Plans for Admission of Women, Personal, February 24, 2016.
7. Hess, Plans for Admission of Women.
8. Hess, Plans for Admission of Women.
9. James P. McCarthy, U.S. Air Force Academy Oral History Interview, Other no. 123, February 18, 1977. Clark Special Collections, USAFA, p. 9.
10. Hess, Plans for Admission of Women.
11. Hess, Plans for Admission of Women.
12. Stanton R. Musser, "Class of 1980 Update: New Training Emphasis," *The Association of Graduates Magazine*, January 1977, p. 10.
13. James R. Allen, U.S. Air Force Oral History Program, Other no. 3, June 21, 1977. Clark Special Collections, USAFA, p. 3.

Chapter 4

1. Henry S. Fellerman, *History of the United States Air Force Academy*. 1 January–31 December 1976 ed. (Colorado Springs: United States Air Force Academy, 1977), p. 298.
2. Grace Lichtenstein, "'Kill, Hate—Mutilate!'" *New York Times Magazine*, September 5, 1976, p. 1.
3. Grace Lichtenstein, "Beauties and the Beast," *Miami News Weekender*, September 18, 1976, p. 3B.
4. Lichtenstein, "'Kill, Hate—Mutilate!'" p. 40.
5. Betsy Muenger, Laurel Scherer, Ken Carter, and Christy Williams, *Women in*

Motion: Celebrating 20 Years at the Air Force Academy (Colorado Springs: U.S. Air Force Academy Directorate of Public Affairs, 1996), p. 15.

6. Jane Seaberry, "Va. Girls Eye AF Careers," *Washington Post*, February 11, 1976, p. B1.

7. Marianne Owens LaRivee, Letter to Kathleen Utley Kornahrens, *Bring Me Men Manuscript*, February 27, 2022.

8. Friends of the Air Force Academy Library, *The Aeronautical History Collection of Colonel Richard Gimbel* (Colorado Springs: Friends of the Air Force Academy Library, n.d.).

9. Lichtenstein, "'Kill, Hate—Mutilate!'" p. 42.

10. Lichtenstein, "'Kill, Hate—Mutilate!'" p. 39.

11. Henry S. Fellerman, *History of the United States Air Force Academy*. 1 January–31 December 1976 ed., Colorado Springs: United States Air Force Academy, 1977, p. 48.

12. Lichtenstein, "'Kill, Hate—Mutilate!'" p. 37.

13. James P. McCarthy, U.S. Air Force Academy Oral History Interview. Other no. 123, February 18, 1977. Clark Special Collections, USAFA, p. 5.

14. James P. McCarthy and Kathleen Utley Kornahrens, General James P. McCarthy Interview, Personal, June 26, 2012.

15. McCarthy, U.S. Air Force Academy Oral History Interview, p. 29.

16. LaRivee, U.S. Air Force Academy Oral History Interview, p. 27.

17. Lichtenstein, "'Kill, Hate—Mutilate!'" p. 42.

Chapter 5

1. George V. Fagan, *The Air Force Academy: An Illustrated History* (Boulder: Johnson Publishing, 1988), p. 53.

2. Fagan, *The Air Force Academy*, p. 53.

3. Henry S. Fellerman, *History of the United States Air Force Academy*, 1 July 1974–30 June 1975 ed. (Colorado Springs: United States Air Force Academy, 1976), p. 106.

4. William J. Wallisch, "The Admission and Integration of Women into the United States Air Force Academy," 1977, p. 195.

5. Fellerman, *History of the United States Air Force Academy*. 1 July 1974–30 June 1975, p. 106.

6. James P. McCarthy, U.S. Air Force Academy Oral History Interview, Other no. 123, February 18, 1977. Clark Special Collections, USAFA, p. 25.

7. McCarthy, U.S. Air Force Academy Oral History Interview, p. 27.

8. McCarthy, U.S. Air Force Academy Oral History Interview, p. 27.

9. James R. Allen, U.S. Air Force Oral History Program, Other no. 3, June 21, 1977, p. 9.

10. Irene Graf and Terry T. Walter, U.S. Air Force Academy Oral History Interview, Other no. 85, September 1, 1977. Clark Special Collections, USAFA, p. 2.

11. Graf and Walter, U.S. Air Force Academy Oral History Interview, p. 48.

12. Graf and Walter, U.S. Air Force Academy Oral History Interview, p. 48.

13. Graf and Walter, U.S. Air Force Academy Oral History Interview, p. 9.

14. Graf and Walter, U.S. Air Force Academy Oral History Interview, p. 11.

15. Graf and Walter, U.S. Air Force Academy Oral History Interview, p. 106.

16. John Ward, John Ward Interview, Personal, March 23, 2022.

17. Graf and Walter, U.S. Air Force Academy Oral History Interview, p. 38.

18. Marianne LaRivee, U.S. Air Force Academy Oral History Interview, Other no. 1106, September 9, 2010. Clark Special Collections, USAFA, p. 7.

Chapter 6

1. June Van Horn Lindner, Letter to Kathleen Utley Kornahrens, *Re: Niner Sisters*, February 20, 2022.

2. Debbie W. Ziebart, Letter to Kathleen Utley Kornahrens, *USAFA Story*, April 7, 2022.

3. Henry S. Fellerman, *History of the United States Air Force Academy*, January–31 December 1976 ed. Colorado Springs: United States Air Force Academy, 1977, p. 36.

4. Lindner, *Re: Niner Sisters*.

5. Jeannette Gaudry Haynie, "Being a WUBA: How Words Shape Power," women.usnagroups.net, December 26, 2018. https://women.usnagroups.net/2018/12/26/-being-a-wuba-how-words-shape-power/.

6. Marianne LaRivee, U.S. Air Force Academy Oral History Interview, Other no. 1106, September 9, 2010. Clark Special Collections, USAFA, p. 7.
7. LaRivee, U.S. Air Force Academy Oral History Interview, p. 14.
8. LaRivee, U.S. Air Force Academy Oral History Interview, p. 12.
9. Gail Colvin, U.S. Air Force Academy Oral History Interview, Other no. 1042, September 15, 2010. Clark Special Collections, USAFA, p. 16.
10. Adam Bernstein, "Robert McDermott; Air Force Academy Dean," *Washington Post*, August 29, 2006. https://www.washingtonpost.com/archive/local/2006/08/29/robert-mcdermott/d878f799-f780-41b8-ba1a-3d32e927a02f/.
11. Karen Wilhelm, Letter to Kathleen Utley Kornahrens, *Publishing*, February 17, 2022.
12. Wilhelm, *Publishing*.
13. Dennis A. Williams and Martin Kasindorf, "The Sexes: Bring Me People," *Newsweek*, July 12, 1976, p. 27.
14. LaRivee, U.S. Air Force Academy Oral History Interview, p. 10.
15. Harry Lalusis, Letter to Kathleen Utley Kornahrens, *Re: USAFA '80. Kathy Utley*, February 28, 2022.
16. Lalusis, *Re: USAFA '80*.
17. James D. Manning, "Memo for Record: Meeting to Discuss OPLAN 7–73," Colorado Springs, USAFA, May 10, 1974.
18. "Commandant's Committee on the Admission of Women Cadets, Minutes Meeting #8," Colorado Springs, USAFA, September 25, 1975, p. 3.
19. James P. McCarthy, U.S. Air Force Academy Oral History Interview, Other no. 123, February 18, 1977. Clark Special Collections, USAFA, p. 27.
20. George V. Fagan, *The Air Force Academy: An Illustrated History* (Boulder: Johnson Publishing, 1988), p. 192.
21. Irene Graf and Terry T. Walter, U.S. Air Force Academy Oral History Interview, Other no. 85, September 1, 1977. Clark Special Collections, USAFA, p. 23.

Chapter 7

1. Lisa Underwood, Letter to Kathleen Utley Kornahrens, *Enjoyed Telling War Stories*, April 5, 2022.
2. Karen Wilhelm, Letter to Kathleen Utley Kornahrens, *Publishing*, February 17, 2022.
3. Jeff Holmquist, "Painting a New Presence," *Checkpoints*, AOG USAFA Alumni Magazine, June 1, 2016, p. 50.
4. Susie Fuselier and Kathleen Utley Kornahrens, Susie Fuselier Interview, Personal, March 7, 2022.
5. Fuselier, Interview.
6. Fuselier, Interview.
7. Fuselier, Interview.
8. Fuselier, Interview.
9. Fuselier, Interview.

Chapter 8

1. "A History of Coeducation," Vassar College Encyclopedia, accessed October 12, 2013. https://www.vassar.edu/vcencyclopedia/notable-events/coeducation/-a-history-of-coeducation.html.
2. Ray Bowden, "Women's History Month: AF Academy's Director of Staff Says Alma Mater Was 'Right Place at the Right Time' • United States Air Force Academy," United States Air Force Academy, March 18, 2022. https://www.usafa.edu/womens-history-month-af-academys-director-of-staff-says-alma-mater-was-right-place-at-the-right-time/.
3. Mark Pimentel, Letter to Kathleen Utley Kornahrens, *Chapel Info ZIP File*, February 25, 2022.
4. Pimentel, *Chapel Info*.
5. Pimentel, *Chapel Info*.
6. George V. Fagan, *The Air Force Academy: An Illustrated History* (Boulder: Johnson Publishing, 1988), p. 81.
7. "Commandant's Committee on the Integration of Women into the Cadet Wing—Minutes Meeting #7" (Colorado Springs: USAFA, September 18, 1975), p. 3.
8. "Ronald L. Sheffield," Washington, D.C., National Air and Space Museum, n.d. https://airandspace.si.edu/support/wall-of-honor/ronald-l-sheffield.
9. June Van Horn Lindner, Letter to Kathleen Utley Kornahrens, *Re: Niner Sisters*, February 20, 2022.

Chapter 9

1. The Air Force Cadet Wing (USAFA), "Air Force Cadet Regulation 50–8, Attachment 3" (Colorado Springs: USAF Academy, December 22, 1976), p. 24.

2. Irene Graf and Terry T. Walter, U.S. Air Force Academy Oral History Interview, Other no. 85, September 1, 1977. Clark Special Collections, USAFA, p. 5.

3. Gene Birkhead, "Women Cadets Fear Femininity Loss," *The Sun*, September 30, 1976.

4. Molly R. Parrish, "Women Cadets Are Feminine: Officer," *Gazette Telegraph*, November 1, 1976, p. 3A.

5. William J. Wallisch, "The Admission and Integration of Women into the United States Air Force Academy," 1977, p. 178.

6. Karen Wilhelm, Letter to Kathleen Utley Kornahrens, *Publishing*, February 17, 2022.

7. Kate Heinzelman, "Yale's First Classes of Women Look Back," *Yale Daily News*, February 20, 2002. https://yaledailynews.com/blog/2002/02/20/yales-first-classes-of-women-look-back/.

8. Marianne LaRivee, U.S. Air Force Academy Oral History Interview, Other no. 1106, September 9, 2010. Clark Special Collections, USAFA, p. 23.

Chapter 10

1. George V. Fagan, *The Air Force Academy: An Illustrated History* (Boulder: Johnson Publishing, 1988), p. 189.

2. *Air Force Academy 1954–2004, 50th Anniversary Oral History* (Colorado Springs: Friends of the Air Force Academy Library, 2005). Clark Special Collections, USAFA, p. 442.

3. J. Seth Bopp, "30 Years of '80s Ladies," United States Air Force Academy, October 13, 2010. https://www.usafa.af.mil/News/Features/Article/429599/30-years-of-80s-ladies/.

4. Marianne LaRivee, U.S. Air Force Academy Oral History Interview, Other no. 1106, September 9, 2010. Clark Special Collections, USAFA, p. 8.

5. Sue Desjardins, Letter to Kathleen Utley Kornahrens, *Sue Desjardin's Review*, February 27, 2022.

Chapter 13

1. June Van Horn Lindner, Letter to Kathleen Utley Kornahrens, *Re: Niner Sisters*, February 20, 2022.

2. "Return with Honor, the Tap Code," PBS, accessed April 25, 2022. https://www.pbs.org/wgbh/americanexperience/features/honor-tap-code/.

Chapter 14

1. Karen Wilhelm, Letter to Kathleen Utley Kornahrens, *Publishing*, February 17, 2022.

Chapter 15

1. Bianca Ruggia, "Mariela Santamaría: Una Mujer con el Poder de Hércules," Ser Argentino—Todo sobre Argentina! April 15, 2021. https://www.serargentino.com/gente/-historias-de-gente/mariela-santamaria-una-mujer-con-el-poder-de-hercules.

Chapter 16

1. Hank Wilborn and Kathleen Utley Kornahrens, Major Hank Wilborn Interview, Personal, March 1, 2022.

2. "Episode 006—Maj. Henry 'Hank' Wilborn (Retired)—the Gabriel Goings Xperiment—Podcast En iVoox." iVoox, April 16, 2021. https://www.ivoox.com/en/-episode-006-maj-henry-hank-wilborn-retired-audios-mp3_rf_80224785_1.html.

3. John Ward, John Ward Interview, Personal, March 23, 2022.

4. Ward, Interview.

5. Ron McNeill, Letter to Kathleen Utley Kornahrens, *Re: Writing*, April 3, 2022.

6. Linda Smith, Linda. "Cadet Had Her Day of Firsts." *Gazette Telegraph*, May 29, 1980.

7. Betsy J. Pimentel, Letter to Kathleen Utley Kornahrens, *Follow-Up*, April 4, 2022.

8. Hans Mark, *An Anxious Peace: A Cold War Memoir* (College Station: Texas A&M University Press, 2019).

9. *USAFA 1980 Graduation Video*, Colorado Springs, USAFA, 1980.

10. William L. Chase and Sarah A. Peterson, "Ready to Take Command," *U.S. News & World Report*, May 26, 1980, p. 36.

11. Jeff Holmquist, "Painting a New Presence," *Checkpoints*, AOG USAFA Alumni Magazine, June 1, 2016, p. 49.

Chapter 17

1. Mark Stevens, "Women Grab Spotlight at AFA Graduation," *Rocky Mountain News*, May 29, 1980, p. 8.

2. Eric Schmitt, "Air Force Academy Zooms in on Sex Cases," *New York Times*, May 1, 1994, p. 1.
3. Schmitt, "Air Force Academy Zooms In," p. 34.
4. Schmitt, "Air Force Academy Zooms In," p. 34.
5. Richard J. Newman, "Upheaval at the Academy," *Air Force Magazine*, January 2004. https://web.archive.org/web/20051114064832/http://afa.org/magazine/jan2004/0104academy.asp.
6. Newman, "Upheaval at the Academy."
7. Newman, "Upheaval at the Academy."
8. "Allegations of Sexual Assault at the U.S. Air Force Academy," Washington, D.C., U.S. Government Printing Office, September 30, 2003. https://www.govinfo.gov/content/pkg/CHRG-108shrg89536/html/CHRG-108shrg89536.htm.
9. Newman, "Upheaval at the Academy."
10. Donna Miles, "Academy Officials: Sexual Assault Reporting Shows System Is Working," American Forces Press Service, January 1, 1970. https://military-online.blogspot.com/2007/12/academy-officials-sexual-assault.html.
11. "Sexual Harassment," U.S. EEOC, accessed May 7, 2022. https://www.eeoc.gov/sexual-harassment.
12. Judith Hicks Stiehm, *Bring Me Men & Women* (Berkeley: University of California Press, 1981), p. 179.
13. Henry S. Fellerman, *History of the United States Air Force Academy*, 1 January–31 December 1976 ed. Colorado Springs: United States Air Force Academy, 1977, p. 118
14. Writer, Times Staff, "'Bring Me Men' Gone: Academy Asks for Ideas," *Tampa Bay Times*, December 14, 2019. https://www.tampabay.com/archive/2003/08/18/bring-me-men-gone-academy-asks-for-ideas/.

Epilogue

1. Debbie W. Ziebart, Letter to Kathleen Utley Kornahrens, *USAFA Story*, April 7, 2022.
2. June Van Horn Lindner, Letter to Kathleen Utley Kornahrens, *Re: Niner Sisters*, February 20, 2022.
3. Marianne Owens LaRivee, Letter to Kathleen Utley Kornahrens, *Bring Me Men Manuscript*, February 27, 2022.
4. *Air Force Academy 1954–2004, 50th Anniversary Oral History* (Colorado Springs: Friends of the Air Force Academy Library, 2005). Clark Special Collections, USAFA, p. 515.
5. *Air Force Academy 1954–2004, 50th Anniversary Oral History*, p. 515.
6. "General Janet C. Wolfenbarger," Air Force: Biography Display, USAF, June 2015. https://www.af.mil/About-Us/Biographies/Display/Article/107934/-general-janet-c-wolfenbarger/.
7. Janet L. Wolfenbarger, Letter to Kathleen Utley Kornahrens, *USAFA Story*, April 7, 2022.
8. Janet L. Wolfenbarger, Letter to Kathleen Utley Kornahrens, *USAFA Story*, April 6, 2022.
9. "Lessons in Leadership, Linda Garcia Cubero," *Latina Style Magazine—National Magazine for the Contemporary Hispanic Woman*, September 2009. https://web.archive.org/web/20090516190434/http://www.latinastyle.com/currentissue/v10-5/punto.html.
10. Bonnie Houchen, Letter to Kathleen Utley Kornahrens, *Condensed Bio—Houchen*, May 7, 2022.
11. Annette Crawford, "Even after Remarkable 32-Year Career, General Continues to Mentor, Inspire," 37th Training Wing, March 28, 2022. https://www.37trw.af.mil/News/Article-Display/Article/2980128/even-after-remarkable-32-year-career-general-continues-to-mentor-inspire/.
12. Crawford, "Even after Remarkable."
13. "AOG Service Spotlight: Gail Colvin: U.S. Air Force Academy AOG & Foundation," August 2021. https://www.usafa.org/Service/SpotlightAugust2021.
14. "AOG Service Spotlight: Gail Colvin: U.S. Air Force Academy AOG & Foundation."
15. Stephen Losey, "Goldfein Unveils Gender-Neutral Update to Air Force Song," *Air Force Times*, February 28, 2020. https://www.airforcetimes.com/news/your-air-force/2020/02/27/goldfein-unveils-gender-neutral-update-to-air-force-song/.
16. Robert Hess, Colonel Robert Hess (Ret.) Plans for Admission of Women, Personal, February 24, 2016.

Bibliography

"Admission of Women into the Cadet Wing, Operations Plan Number 76–75." Colorado Springs: USAFA, October 7, 1975.

"Admission of Women into the Cadet Wing, Operations Plan Number 76–75, Annex C." Colorado Springs: USAFA, October 7, 1975.

"Admission of Women into the Cadet Wing, Operations Plan Number 76–75, Appendix IV to Annex D, Basic Cadet Training Program (BCT)." Colorado Springs: USAFA, October 7, 1975.

"Admission of Women into the Cadet Wing, Operations Plan Number 76–75, Appendix IV to Annex D, Intercollegiate Athletics." Colorado Springs: USAFA, October 7, 1975.

"Admission of Women into the Cadet Wing, Operations Plan Number 76–75, Appendix IV to Annex D, Intramural Program." Colorado Springs: USAFA, October 7, 1975.

"Admission of Women into the Cadet Wing, Operations Plan Number 76–75, Appendix IV to Annex D, Physical Education Instructional Program." Colorado Springs: USAFA, October 7, 1975.

Air Force Academy 1954–2004, 50th Anniversary Oral History. Colorado Springs: Friends of the Air Force Academy Library, 2005. Clark Special Collections, USAFA.

"Air Force Academy Sexual Misconduct Investigation." CBS News. CBS Interactive, December 10, 2017. https://www.cbsnews.com/video/playlist/air-force-academy-sexual-misconduct-investigation/.

The Air Force Cadet Wing (USAFA). "Air Force Cadet Regulation 50–8, Attachment 3." Colorado Springs: USAF Academy, December 22, 1976.

"Allegations of Sexual Assault at the U.S. Air Force Academy." Washington, D.C.: U.S. Government Printing Office, September 30, 2003. https://www.govinfo.gov/content/pkg/CHRG-108shrg89536/html/CHRG-108shrg89536.htm.

Allen, James R. U.S. Air Force Oral History Program. Other no. 3, June 21, 1977. Clark Special Collections, USAFA.

"Annual Reports on Sexual Assault at the Military Service Academies, 2005–2006 Academic Year." www.sapr.mil/public/docs/reports. Defense Manpower Data Center (DMDC), December 2006. https://www.sapr.mil/public/docs/reports/APY_05-06_MSA_Report.pdf.

"AOG Service Spotlight: Gail Colvin: U.S. Air Force Academy AOG & Foundation." August 2021. https://www.usafa.org/Service/SpotlightAugust2021.

"Appendix I to Annex G to OPLAN 36–72, Time Frame Input Requirements." Colorado Springs: USAFA, September 15, 1972.

"Appendix I to Annex I to OPLAN 36–72, Reply to Query." Colorado Springs: USAFA, September 15, 1972.

Associated Press. "Women at 4 Service Academies Viewed as Equals to Men Cadets." *New York Times*, September 11, 1977.

Baillie, Amber. "'What Women Can Do': Female Air Training Officers Tore down Walls." United States Air Force Academy: Features. Academy Spirit, March 15, 2013. https://www.usafa.af.mil/News/Features/Article/429475/what-women-can-do-female-air-training-officers-tore-down-walls/.

Bibliography

BEAST '76, Ready-Halt. 1. Class of 1980 ed. Vol. 1. USAFA, 1976.

Beck, Stanley C., and Walter R. Berg. U.S. Air Force Oral History Program. Other no. 422, January 14, 1988. Clark Special Collections, USAFA.

Bernstein, Adam. "Robert McDermott; Air Force Academy Dean." *Washington Post,* August 29, 2006. https://www.washingtonpost.com/archive/local/2006/08/29/robert-mcdermott/d878f799-f780-41b8-ba1a-3d32e927a02f/.

Birkhead, Gene. "Women Cadets Fear Femininity Loss." *The Sun,* September 30, 1976.

Bopp, J. Seth. "30 Years of '80s Ladies." United States Air Force Academy, October 13, 2010. https://www.usafa.af.mil/News/Features/Article/429599/30-years-of-80s-ladies/.

Bowden, Ray. "Women's History Month: AF Academy's Director of Staff Says Alma Mater Was 'Right Place at the Right Time' • United States Air Force Academy." United States Air Force Academy, March 18, 2022. https://www.usafa.edu/womens-history-month-af-academys-director-of-staff-says-alma-mater-was-right-place-at-the-right-time/.

Campbell, Kathleen. Letter to Kathleen Utley Kornahrens. *Re: Question to 80's Ladies,* March 17, 2003.

Chase, William L., and Sarah A. Peterson. "Ready to Take Command." *U.S. News & World Report,* May 26, 1980.

Clark, A.P. Letter to Lt General Robert J. Dixon. Colorado Springs: U.S. Air Force Academy, April 9, 1973.

Colvin, Gail. U.S. Air Force Academy Oral History Interview. Other no. 1042, September 15, 2010. Clark Special Collections, USAFA.

"The Coming American by Sam Walter Foss." The Bard on the Hill, December 2, 2011. https://thebardonthehill.wordpress.com/2011/12/02/the-coming-american-by-sam-walter-foss/.

"Commandant's Committee on the Admission of Women Cadets, Minutes Meeting #4." Colorado Springs: USAFA, September 4, 1975.

"Commandant's Committee on the Admission of Women Cadets, Minutes Meeting #8." Colorado Springs: USAFA, September 25, 1975.

"Commandant's Committee on the Admission of Women Cadets, Minutes Meeting #11." Colorado Springs: USAFA, November 3, 1975.

"Commandant's Committee on the Admission of Women Cadets, Minutes Meeting #16." Colorado Springs: USAFA, February 2, 1976.

"Commandant's Committee on the Integration of Women Cadets, Minutes Meeting #5." Colorado Springs: USAFA, September 5, 1975.

"Commandant's Committee on the Integration of Women into the Cadet Wing—Minutes Meeting #7." Colorado Springs: USAFA, September 18, 1975.

Conley, Kathleen. Letter to Kathleen Kornahrens. *Re: USAFA Writing,* April 19, 2022.

Contrails: The Air Force Cadet Handbook 22, 1976–1977 ed. Colorado Springs: United States Air Force Academy, 1976.

Cooper, Robert R. U.S. Air Force Academy Oral History Interview. Other no. 1107, September 22, 2010. Clark Special Collections, USAFA.

Crawford, Annette. "Even after Remarkable 32-Year Career, General Continues to Mentor, Inspire." 37th Training Wing, March 28, 2022. https://www.37trw.af.mil/News/Article-Display/Article/2980128/even-after-remarkable-32-year-career-general-continues-to-mentor-inspire/.

Cubero, Linda Garcia. Letter to Kathleen Utley Kornahrens. *Questions to 80's Ladies,* March 17, 2003.

Desjardins, Sue. Letter to Kathleen Utley Kornahrens. *Sue Desjardin's Review,* February 27, 2022.

Directorate of Plans and Programs. "Admission Criteria for Selection of the Class of 1980." Colorado Springs: USAFA, December 11, 1975.

Directorate of Plans and Programs. *Operations Plan, Admission of Women into the Cadet Wing.* Number 76–75 ed. Colorado Springs: USAF Academy, 1975.

"ECFR: 34 CFR Part 106—Nondiscrimination on the Basis...." eCFR. National Archives and Records Administration. Accessed April 25, 2022. https://www.ecfr.gov/current/title-34/subtitle-B/chapter-I/part-106.

The Editors of Encyclopedia Britannica.

"WAVES." *Encyclopedia Britannica*, September 27, 2013. https://www.britannica.com/topic/WAVES-United-States-naval-organization.

"Episode 006—Maj. Henry 'Hank' Wilborn (Retired)—the Gabriel Goings Xperiment—Podcast En iVoox." iVoox, April 16, 2021. https://www.ivoox.com/en/episode-006-maj-henry-hank-wilborn-retired-audios-mp3_rf_80224785_1.html.

"Exemptions from Title IX." Home. U.S. Department of Education (ED), March 8, 2021. https://www2.ed.gov/about/offices/list/ocr/docs/t9-rel-exempt/index.html.

Fagan, George V. *The Air Force Academy: An Illustrated History*. Boulder: Johnson Publishing, 1988.

Farrell, Kate. Letter to Kathleen Utley Kornahrens. Re: *80's Ladies Group Letter to USAFA*, April 28, 2003.

Fellerman, Henry S. *History of the United States Air Force Academy*. 1 January–31 December 1976 ed. Colorado Springs: United States Air Force Academy, 1977.

Fellerman, Henry S. *History of the United States Air Force Academy*. 1 January–31 December 1977 ed. Colorado Springs: United States Air Force Academy, 1978.

Fellerman, Henry S. *History of the United States Air Force Academy*. 1 January–31 December 1978 ed. Colorado Springs: United States Air Force Academy, 1979.

Fellerman, Henry S. *History of the United States Air Force Academy*. 1 January–31 December 1979 ed. Colorado Springs: United States Air Force Academy, 1980.

Fellerman, Henry S. *History of the United States Air Force Academy*. 1 January–31 December 1980 ed. Colorado Springs: United States Air Force Academy, 1981.

Fellerman, Henry S. *History of the United States Air Force Academy*. 1 July 1974–30 June 1975 ed. Colorado Springs: United States Air Force Academy, 1976.

Fellerman, Henry S. *History of the United States Air Force Academy*. 1 July–31 December 1975 ed. Colorado Springs: United States Air Force Academy, 1976.

Fellerman, Henry S. *History of the United States Air Force Academy*. 1 July 1972–30 June 1973 ed. Colorado Springs: United State Air Force Academy, 1973.

"The Final Report of the Panel to Review Sexual Misconduct Allegations at the U.S. Air Force Academy: Hearing before the Total Force Subcommittee of the Committee on Armed Services, House of Representatives, One Hundred Eighth Congress, First Session, Hearing Held September 24, 2003: United States. Congress. House. Committee on Armed Services. Subcommittee on Total Force: Free Download, Borrow, and Streaming." Internet Archive. Washington, D.C.: U.S. GPO: For sale by the Superintendent of Documents, U.S. GPO, September 24, 2003. https://archive.org/details/finalreportofpan00unit/mode/2up?q=%22one%2Bis%2Btoo%2Bmany%22.

"Final Report on Academy Sexual Misconduct." C-SPAN, September 23, 2003. https://www.c-span.org/video/?178311-4%2Ffinal-report-academy-sexual-misconduct#.

Fraser, Donald M., Fortney H. Scott, and Jacqueline Cochran. *Hearing Before Armed Services Committee, Concerning the Admission of Women to the Service Academies*. H.R. 10705 ed. Washington, D.C.: U.S. Congress, 1974.

Friends of the Air Force Academy Library. *The Aeronautical History Collection of Colonel Richard Gimbel*. Colorado Springs: Friends of the Air Force Academy Library, n.d.

Fuselier, Susie, and Kathleen Utley Kornahrens. Susie Fuselier Interview, Personal, March 7, 2022.

Garcia, Linda. Letter to Kathleen Utley Kornahrens. *USAFA Writing*, April 7, 2022.

Garvin, Honi J. Letter to Kathleen Utley Kornahrens. *Joanie French*, April 10, 2003.

Garvin, Honi J., and Kathleen Utley Kornahrens. Honi Garvin Interview, Personal, October 1, 2003.

Gebecki, Mark E. Air Force Academy Gender and Racial Disparities: Report to Congressional Requesters § (1993).

Gebecki, Mark E. DOD Service Academies: More Actions Needed to Eliminate Sexual Harassment: Report to Congressional Requesters § (1994).

"General Janet C. Wolfenbarger." Air Force: Biography Display. USAF, June 2015. https://www.af.mil/About-Us/Biographies/Display/Article/107934/-general-janet-c-wolfenbarger/.

Graf, Irene, and Terry T. Walter. U.S. Air Force Academy Oral History Interview. Other no. 85, September 1, 1977. Clark Special Collections, USAFA.

Haynie, Jeannette Gaudry. "Being a WUBA: How Words Shape Power." women.usnagroups.net, December 26, 2018. https://women.usnagroups.net/2018/12/26/being-a-wuba-how-words-shape-power/.

Heinzelman, Kate. "Yale's First Classes of Women Look Back." *Yale Daily News*, February 20, 2002. https://yaledailynews.com/blog/2002/02/20/yales-first-classes-of-women-look-back/.

Hess, Robert. Colonel Robert Hess (Ret.) Plans for Admission of Women, Personal, February 24, 2016.

Hess, Robert. Letter to Kathleen Utley Kornahrens, *Re: Women at USAFA*, February 2, 2016.

"History: United States Air Force Academy." United States Air Force Academy, February 26, 2022. https://www.usafa.edu/about/history/.

"A History of Coeducation." Vassar College Encyclopedia. Accessed October 12, 2013. https://www.vassar.edu/vcencyclopedia/notable-events/coeducation/a-history-of-coeducation.html.

Holmquist, Jeff. "Painting a New Presence." *Checkpoints*, AOG USAFA Alumni Magazine, June 1, 2016.

Houchen, Bonnie. Letter to Kathleen Utley Kornahrens. *Condensed Bio—Houchen*, May 7, 2022.

Janofsky, Michael. "Air Force Begins an Inquiry of Ex-Cadets' Rape Charges." *New York Times*, February 20, 2003. https://www.nytimes.com/2003/02/20/us/air-force-begins-an-inquiry-of-ex-cadets-rape-charges.html?searchResultPosition=1.

Janofsky, Michael. "Top Air Force Officer, at Academy, Issues Warning." *New York Times*, March 8, 2003. https://www.nytimes.com/2003/03/08/us/top-air-force-officer-at-academy-issues-warning.html?searchResultPosition=1#.

Johnson, Kimberly. "When Women Earned Their Wings: The USAF's First Crop of Female Pilots." *Flying Magazine*, March 30, 2022. https://www.flyingmag.com/when-women-earned-their-wings-the-usafs-first-crop-of-female-pilots/.

Juhas, Diane. Letter to Kathleen Utley Kornahrens. *Update to the 80's Ladies*, March 11, 2003.

Kenworth, Tom, and Patrick O'Driscoll. "Climate Has to Change, Air Force Leader Says." *USA Today*, March 13, 2003. http://www.usatoday.com/news/nation/2003-03-12-academy_x.htm.

Lalusis, Harry. Letter to Kathleen Utley Kornahrens. *Re: USAFA '80. Kathy Utley*, February 28, 2022.

LaRivee, Marianne. U.S. Air Force Academy Oral History Interview. Other no. 1106, September 9, 2010. Clark Special Collections, USAFA.

LaRivee, Marianne Owens. Letter to Kathleen Utley Kornahrens. *Bring Me Men Manuscript*, February 27, 2022.

"Lessons in Leadership, Linda Garcia Cubero." *Latina Style Magazine—National Magazine for the Contemporary Hispanic Woman*, September 2009. https://web.archive.org/web/20090516190434/http://www.latinastyle.com/currentissue/v10-5/punto.html.

Lichtenstein, Grace. "Beauties and the Beast." *Miami News Weekender*, September 18, 1976.

Lichtenstein, Grace. "'Kill, Hate—Mutilate!'" *New York Times Magazine*, September 5, 1976.

Lichtenstein, Grace. "A Year Later How Women Are Faring at the Air Force Academy." *New York Times Magazine*, September 11, 1977.

Lichtenstein, Grace. "Sex Barrier Falls as Women Enter Air Force Academy." *New York Times*, June 29, 1976, Special. https://www.nytimes.com/1976/06/29/archives/sex-barrier-falls-as-women-enter-air-force-academy.html.

Lindner, June Van Horn. Letter to Kathleen Utley Kornahrens. *Re: Niner Sisters*, February 20, 2022.

Losey, Stephen. "Goldfein Unveils Gender-Neutral Update to Air Force Song." *Air Force Times*, February 28, 2020. https://www.airforcetimes.com/news/your-air-force/2020/02/27/goldfein-unveils-gender-neutral-update-to-air-force-song/.

Manning, James D. "Memo for Record: Meeting to Discuss OPLAN 7-73." Colorado Springs: USAFA, May 10, 1974.

Mark, Hans. *An Anxious Peace: A Cold*

War Memoir. College Station: Texas A&M University Press, 2019.
McCarthy, James P. U.S. Air Force Academy Oral History Interview. Other no. 123, February 18, 1977. Clark Special Collections, USAFA.
McCarthy, James P., and Kathleen Utley Kornahrens. General James P. McCarthy Interview, Personal, June 26, 2012.
McNeill, Ron. Letter to Kathleen Utley Kornahrens. *Re: Writing*, April 3, 2022.
Miles, Donna. "Academy Officials: Sexual Assault Reporting Shows System Is Working." American Forces Press Service, January 1, 1970. https://military-online.blogspot.com/2007/12/academy-officials-sexual-assault.html.
Miller, Ed Mack. *Wild Blue U: The Story of the U.S. Air Force Academy*. New York: Macmillan, 1972.
Mueller, Jack. *Core Values to Be New Words above the Warrior Ramp*. September 23, 2004.
Muenger, Betsy, Laurel Scherer, Ken Carter, and Christy Williams. *Women in Motion: Celebrating 20 Years at the Air Force Academy*. Colorado Springs: U.S. Air Force Academy Directorate of Public Affairs, 1996.
Muenger, E.A. "Fact Sheet on the Integration of Women at the Air Force Academy." Edited by Joan Varra. Colorado Springs: United States Air Force Academy, March 1977.
Musser, Stanton R. "Class of 1980 Update: New Training Emphasis." *The Association of Graduates Magazine,* January 1977.
Nellenbach, Joanita. "Air Force Cadet Utley Finishes 4 Busy Years." *Fort Meade Ledger,* March 25, 1980.
Newman, Richard J. "Upheaval at the Academy." *Air Force Magazine,* January 2004. https://web.archive.org/web/20051114064832/http://afa.org/magazine/jan2004/0104academy.asp.
"News Release, United States Air Force, U.S. Air Force Academy." October 8, 1975.
Parrish, Molly R. "Women Cadets Are Feminine: Officer." *Gazette Telegraph,* November 1, 1976.
Pimentel, Betsy J. Letter to Kathleen Utley Kornahrens. *Follow-Up*, April 4, 2022.
Pimentel, Betsy Joviak. Letter to Kathleen Utley Kornahrens. *Re: Publishing*, February 22, 2022.
Pimentel, Mark. Letter to Kathleen Utley Kornahrens. *Chapel Info ZIP File*, February 25, 2022.
Playboy editors. "Our Strong Suit." *Playboy,* December 17, 2019.
Regan, Tanya. Letter to Kathleen Utley Kornahrens. *Re: Question to 80's Ladies*, March 17, 2003.
"Return with Honor, the Tap Code." PBS. Accessed April 25, 2022. https://www.pbs.org/wgbh/americanexperience/features/honor-tap-code/.
Roche, James G. "Speech to National Character and Leadership Symposium." 10th Annual National Character and Leadership Symposium, February 27, 2003.
Roeder, Tom. "A Broken Code: Air Force Academy Athletes Flouted Sacred Honor Code by Committing Sexual Assaults, Taking Drugs, Cheating and Engaging in Other Misconduct." *Colorado Springs Gazette,* August 2, 2014. https://gazette.com/sports/a-broken-code/article_97b9e453-fcab-5ee7-a577-7743c03260f3.html.
"Ronald L. Sheffield." Washington, D.C.: National Air and Space Museum, n.d. https://airandspace.si.edu/support/wall-of-honor/ronald-l-sheffield.
Ruggia, Bianca. "Mariela Santamaría: Una Mujer con el Poder de Hércules." Ser Argentino—Todo sobre Argentina! April 15, 2021. https://www.serargentino.com/gente/historias-de-gente/mariela-santamaria-una-mujer-con-el-poder-de-hercules.
Schemo, Diana Jean. "Academy Cadet Chief Backs Rape Report Disclosures." *New York Times,* July 17, 2003. https://www.nytimes.com/2003/07/17/us/academy-cadet-chief-backs-rape-report-disclosures.html?searchResultPosition=1#.
Schmitt, Eric. "Air Force Academy Zooms in on Sex Cases." *New York Times,* May 1, 1994.
Schmitt, Eric. "Top Air Force General Backs Independent Inquiry in Rapes." *New York Times,* February 27, 2003. https://www.nytimes.com/2003/02/27/us/top-air-force-general-backs-independent-inquiry-in-rapes.html?searchResultPosition=2#.

Seaberry, Jane. "Va. Girls Eye AF Careers." *Washington Post*, February 11, 1976.

"Sex Discrimination: Overview of the Law." U.S. Department of Education (ED), February 16, 2022. https://www2.ed.gov/policy/rights/guid/ocr/sexoverview.html.

"Sexual Harassment." U.S. EEOC. Accessed May 7, 2022. https://www.eeoc.gov/sexual-harassment.

Sheffield, Ronald L., and Kathleen Utley Kornahrens. Ron Sheffield Interview, Personal, March 16, 2022.

Smith, Linda. "Cadet Had Her Day of Firsts." *Gazette Telegraph*, May 29, 1980.

Solanes, Gustavo. "Una Mujer en el Aire." *Mendoza*, July 24, 1979.

Stevens, Mark. "Women Grab Spotlight at AFA Graduation." *Rocky Mountain News*, May 29, 1980.

Stiehm, Judith Hicks. *Bring Me Men & Women*. Berkeley: University of California Press, 1981.

Taylor, John B. "The New Cadets." *Airman* XX, no. 12, December 1976.

Thornhill, Paula. U.S. Air Force Academy Oral History Interview. Other no. 270, April 30, 1980. Clark Special Collections, USAFA.

"Title IX and Sex Discrimination." U.S. Department of Education (ED), August 20, 2021. https://www2.ed.gov/about/offices/list/ocr/docs/tix_dis.html.

Underwood, Lisa. Letter to Kathleen Utley Kornahrens. *Enjoyed Telling War Stories*, April 5, 2022.

Underwood, Lisa. Letter to Kathleen Utley Kornahrens. *"Shady Rest" Re: Question to 80's Ladies*, March 18, 2003.

United States Air Force Academy. *USAFA Catalog 1975-1976, Addendum for Women Admissions Procedures*. Colorado Springs: USAFA, 1975.

"USAFA Contingency Plan Number 36-72, Integration of Females into the Cadet Wing, Annex A." Colorado Springs: USAFA, September 15, 1972.

"USAFA Contingency Plan Number 36-72, Integration of Females into the Cadet Wing, Annex G." Colorado Springs: USAFA, July 1972.

"USAFA Contingency Plan Number 36-72, Integration of Females into the Cadet Wing, Fake Cover." Colorado Springs: USAFA, September 15, 1972.

USAFA 1980 Graduation Video. Colorado Springs: USAFA, 1980.

Wallisch, William J. "The Admission and Integration of Women into the United States Air Force Academy." Dissertation, University of Southern California, 1977.

Ward, John. John Ward Interview, Personal, March 23, 2022.

Ward, John. Letter to Kathleen Utley Kornahrens. *USAFA Story*, March 28, 2022.

Watson, Gary. "Peter, Paul and Mary Tell Jone French Story." *The Free Press Second Front Page*, April 13, 1970.

Webguy2.0, USAFA Web Team. "Webguy's Blog." Basic Jacks Valley Info | USAFA Webguy, July 24, 2017. https://usafawebguy.com/Blog/Entry/1889.

Wilborn, Hank, and Kathleen Utley Kornahrens. Major Hank Wilborn Interview, Personal, March 1, 2022.

Wilhelm, Karen. Letter to Kathleen Utley Kornahrens. *Publishing*, February 17, 2022.

Williams, Dennis A., and Martin Kasindorf. "The Sexes: Bring Me People." *Newsweek*, July 12, 1976.

Wolfenbarger, Janet L. Letter to Kathleen Utley Kornahrens. *USAFA Story*, April 6, 2022.

Wolfenbarger, Janet L. Letter to Kathleen Utley Kornahrens. *USAFA Story*, April 7, 2022.

Women's Research and Education Institute. "Chronology of Significant Legal and Policy Changes Affecting Women in the Military: 1947–2003." Arlington, VA, n.d.

Writer, Times Staff. "'Bring Me Men' Gone: Academy Asks for Ideas." *Tampa Bay Times*, December 14, 2019. https://www.tampabay.com/archive/2003/08/18/-bring-me-men-gone-academy-asks-for-ideas/.

Ziebart, Debbie W. Letter to Kathleen Utley Kornahrens. *USAFA Story*, April 7, 2022.

Index

Numbers in ***bold italics*** indicate pages with illustrations

Allen, Gen. James R. 36, 56, 88–89, 93
ATOs 40–46, 51, 52, 68, 69, 79, 84–91, 103–106, 119, 136, 137, 141, 147, 152–155, 186, 202, 228, 231

Clark, Gen. A.P. 51, 148
Contrails ***34***, 35, ***36***, 55, 59, 82, 93, 94, 98–100

Galloway, Capt. Judith 105, 141
Gathwright, Lt. Paula 86, 119
Graf, Lt. Irene 86, 88

Hell Week 117, 155–161
Hess, Col. Robert 51, 53, 85, 231

inprocessing ***29***, 33, 34, 45, 114, 152
integration 21, 31, 35, 51, 53, 54, 79, 101, 103, 106, 137, 141, 148, 149, 218, 228

Mark, Dr. Hans 208
McCarthy, Gen. James P. 32, 37, 67–69, 87, 88

press 15, 32, 37, 58, 64, 65, 87, 145, 206, 208–210, 216

SERE 162–164, 168–173, 176, 178, 179, 205, 210
spirit missions 7, 126, 127, 132, 148, 150, 152

Walter, Terry T. Lt. 87–90, 106, 137
Women's Area 3, 40, 45, 93, 103–105, 127, 147–149, 152, 154, 160, 177

245

www.ingramcontent.com/pod-product-compliance
Ingram Content Group UK Ltd.
Pitfield, Milton Keynes, MK11 3LW, UK
UKHW041937140426
5217IPUK00014B/526